D0242175

9.99

Genetics

A concise introduction for students

5728

'07

THE COLLEGE OF
RICHARD COLLYER
LIBRARY

Studymates

Many other titles in preparation

Genetics

A concise introduction for students

Edwin Oxlade

BA DPhil CertEd

www.studymates.co.uk

© Copyright 1999 by Edwin Oxlade

First published in 1999 by Studymates, a Division of International Briefings Ltd, Plymbridge House, Estover Road, Plymouth PL6 7PY, United Kingdom.

Telephone: (01752) 202301.
Fax: (01752) 202333

Email (Publisher):   publisher@studymates.co.uk
Email (Customer Services):   cservs@plymbridge.com
Series web site: http://www.studymates.co.uk

All rights reserved. No part of this work may be reproduced or stored in an information retrieval system without the express permission of the Publishers given in writing.

Note: The contents of this book are offered for the purposes of general guidance only and no liability can be accepted for any loss or expense incurred as a result of relying in particular circumstances on statements made in this book. Readers are advised to check the current position with the appropriate authorities before entering into personal arrangements.

Printed and bound by The Cromwell Press Ltd, Trowbridge, Wiltshire.
Typeset by PDQ Typesetting, Newcastle-under-Lyme, Staffordshire.

Contents

List of Illustrations

Preface

This book aims to cover all those aspects of genetics that may be included in school and introductory college and university courses. It assumes no prior background in the subject and concentrates on covering the necessary basic knowledge and the main concepts. Inevitably, in a book of this length, some detail is omitted. It is hoped, however, that the advantages of limiting the content and trying to keep the approach simple justify the end result. Comprehensive books on genetics, these days, tend to be very large, very expensive and are a daunting prospect for anyone, especially someone tackling this fascinating subject for the first time.

The book has been written with the reader in mind. There is a need for a user-friendly genetics text that covers everything from the founding experiments of Mendel to the most recent work in genetic technology, giving equal weight to both classical and molecular genetics and avoiding unnecessary terminology and particularly difficult or specialised areas. It is designed as a supplementary resource for a student embarking on a genetics course, something to give, perhaps, a slightly different perspective on topics and an overall view of the subject, to fill in gaps, to facilitate and reinforce learning and to act as a revision aid. Although there is a logic behind the sequence of chapters and every topic does require some understanding of other areas of the subject, the book can quite easily be used selectively and the chapters covered in any order.

If any part of the book makes the learning and understanding of genetics easier, or gives new insights into this large and complex subject, it will have achieved its aim.

Edwin Oxlade

edwinoxlade@studymates.co.uk

1

Genetics in Context

One minute summary – Cells and organisms are constantly reproducing themselves. Genetics is concerned with the way in which, during reproduction, continuity of biological characteristics is ensured by the passing of information from one generation to the next. Genetic information is found in the universal material of inheritance, the nucleic acid, DNA. The ability of DNA to replicate accurately means that the mechanism of inheritance is a conservative one. New cells and organisms are either genetically identical to those from which they came or are similar to a very high degree. Some variation, however, is vital to the long term survival of species. This comes about through the combination and reassortment of genetic material from different sources, during the sexual process. In this chapter you will learn about:

▶ the meaning of genetics
▶ the simplest form of inheritance
▶ asexual reproduction
▶ the sex process
▶ continuity and change in heredity
▶ the requirements of a genetic material
▶ a brief history of genetics

What is genetics?

1. Genetics is the study of the inheritance of biological characteristics. Whenever a cell divides to form new cells, or a living organism reproduces, the new generation is supplied with information from its parent or parents. We call this information genetic information. It is, ultimately, what determines the characteristics of the new cells or individuals.

11

2. Genetics is also the study of the genetic information itself and the way in which it functions at a molecular level.

To distinguish between the two approaches to genetics, the study of the inheritance of observable characteristics and the biochemical approach, we refer to classical genetics and molecular genetics, respectively.

Genotype and phenotype

Genetic information and inherited characteristics are not the same thing. The observable character of a cell or organism is known as its phenotype. This is the expression of the genetic information. The information itself, the genetic makeup of a cell or organism, is its genotype.

It is important to appreciate that genetic information is no more than a plan or blue print which carries the potential for any characteristic that might appear in the phenotype. It is not the characteristics themselves.

Genotype is the major determinant of phenotype, but environmental and developmental factors have a powerful modifying influence. A certain genotype might only be expressed, for example, under particular environmental conditions, or at one stage in the life cycle or, in the case of a multicellular organism, in some of the cells but not in others.

As a general rule: phenotype is good evidence of genotype but genotype may not be a predictor of phenotype. A bald man, for example, *must* carry genetic information for baldness. A woman or a young boy can carry the same information and not be bald.

The basic tenet of genetics

At the root of all genetics is the fundamental idea that biological characteristics are determined by a molecular plan which is passed from cell to cell and from generation to generation according to mechanisms that are the same for all living organisms.

We now know that the common material for inheritance in all living things is DNA. Modern genetics, therefore, is, to a large extent, concerned with the study of DNA and DNA technology. So central is DNA to genetics that the whole of the next chapter is devoted to the subject.

Different fields of genetic study

Because genetics is a central theme running through all biology, several fields of study contribute to it. Fig. 1 is an attempt to show how different scientific disciplines have contributed and still contribute to the advance of genetics.

The simplest form of inheritance

Inheritance need not involve a male and female side. It doesn't have to involve whole organisms. Any passing of genetic information from one biological entity to another counts as inheritance. So the simplest form of inheritance is possibly when one cell divides to form two.

(a) The cell makes a copy of its genetic material.

(b) The two identical pieces of genetic information are separated and apportioned, one to each side of the cell.

(c) The cell divides.

(d) There are now two cells which carry the same genetic information. Each new cell has inherited the genetic type of the original cell.

Differentiation

Two cells with the same genetic material are not necessarily the same. During differentiation of cells in a multicellular organism, cells with the same genetic information may well turn out to be very different. Differentiation is the selective expression of parts of the genetic material of different cells so that a variety of cell types is generated from a single set of genetic instructions.

One of the unresolved problems in genetics is how only a small part of the genetic information of a cell is ever activated and how the part that is activated is different in different cells and at different times.

Totipotency

Looking at this the other way suggests that very different cells of a single organism can be genetically identical. Does a human gut cell contain the same genetic information as a bone cell or a skin cell or, indeed any other cell of the body? The answer, in many cases, is yes.

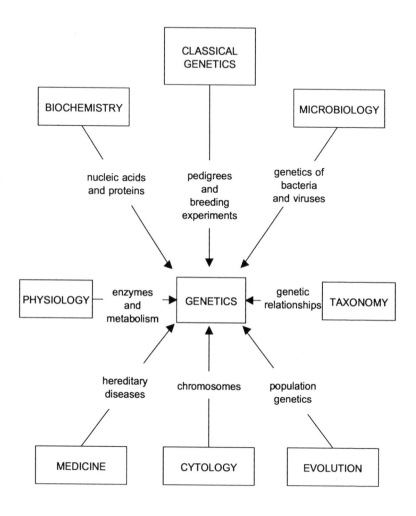

Fig. 1. Fields of scientific study and their relationship to genetics.

After all, every cell of the body is derived from one fertilised egg cell by a process of many cell divisions, at each of which the genetic material of the cell is replicated.

The idea that every cell of a multicellular organism contains *all* the genetic information needed for development of the whole organism is known as totipotency. There are at least three pieces of evidence for the idea.

Evidence for totipotency

(a) If an organism has grown from a single cell, whether it was a fertilised egg or a spore or any other type of single cell, that cell *must* have contained all the information needed to produce the whole organism. And every cell of the organism has come from that one original cell by cell division. Cell division is a process which is expected to copy the genetic material exactly.

(b) Whole plants can be regenerated from single cells taken from differentiated tissues of the plant. For example, carrot plants have been grown from single cells originating in stem, root, petiole and embryo tissue. These plants are normal in every respect and able to reproduce by seed.

(c) In animal cloning the genetic information in an egg cell is removed and replaced with genetic material taken either from a cell of the same animal or from a cell of an animal of the same type. The 'egg' proceeds to develop normally into an adult which is, effectively, the twin of the animal which donated its genetic material.

Cell division and inheritance – further evidence

Cloning is not only evidence that a differentiated cell of the body contains the animal's complete genetic complement. It also shows that the genetic information is inherited in a predictable and planned way at every cell division that takes place during the growth and development of the cloned animal. How else could it grow into a twin of the donor?

The natural occurrence of identical twins is further evidence that the inheritance that takes place at cell division involves the passing on of two identical copies of the genetic information to each daughter cell. Identical twins are produced when the fertilised egg (the zygote) divides into two and each cell then develops into a separate individual. This is why identical twins are referred to as monozygotic twins.

Asexual reproduction

Many single celled organisms reproduce by cell division. One cell becomes two. Two become four and so on. This type of reproduction is asexual. Each new individual inherits the exact genotype of the original.

All asexual reproduction, in multicellular as well as unicellular organisms, gives rise to genetically identical progeny because it only involves replicative cell division.

The sex process

Sex is the process by which genetic variety is generated. The whole of classical genetics has been, in effect, a search for the genetic consequences of sex. From a genetic viewpoint, there are three essential features to a sexual process:

1. The joining together of two separate pieces of genetic information. Usually one is from the 'male' side and one from the 'female'.

2. At some point in the life cycle, a reduction in the amount of genetic information per cell. Since the joining together of two lots of genetic material increases the amount, something has to bring it back to what it was before.

3. A shuffling or recombination of different parts of the genetic information, leading to genetic variety.

Sex is usually associated with reproduction but, strictly, the sexual process, by itself, is not a reproductive event.

The sexual cycle
How does sex fit into a typical animal life cycle? We start with a sperm and an egg. The sperm contains one set of genetic information and the egg contains another. Except in the case of a hermaphrodite

animal fertilising itself, the sperm and the egg have come from different individuals.

When the sperm and egg fuse in the act of fertilisation a single cell called a zygote is formed. The two sets of genetic information combine in the zygote which now has twice as much genetic material as either the sperm or the egg. As the zygote develops to form a new individual every one of this new individual's cells will also have two sets of genetic material.

When the new individual comes to make sperms or eggs, or both, there is a problem. From cells with two lots of genetic material it has to make gametes each containing just one set of genetic material. To do this it requires a special type of cell division known as reduction division or meiosis. Reduction division halves the amount of genetic material in each cell so that we get back to sperms and eggs with only one set of genetic material – which is where we started (fig. 2).

How sex generates variety

Later chapters will cover this topic in more detail. It is important now to appreciate that, if it were not for the fact that organisms include a sexual phase in their life cycles, there would be little or no differences between individuals, and genetics would be a much less complex subject. Virtually the whole of classical genetics, including the pioneering work of Gregor Mendel would never have happened. The whole basis of most genetic research has been to interpret the changes brought about by sexual processes.

Three factors in a sexual cycle generate variety:

▶ *The segregation of genetic material into gametes – or into cells that later produce gametes.*
Gametes contain half the genetic material of a cell that produced them. The question is ....which half? Without going into the details, which you will find elsewhere, when gametes are produced by meiosis, every gamete contains a slightly different half. Every gamete, therefore is, more likely than not, genetically unique. The reason for this is that there is a random element to the distribution of half the genetic material into each gamete.

▶ *The fusion of gametes* – The fusion of gametes is also random. In animals huge numbers of sperm are produced, any one of

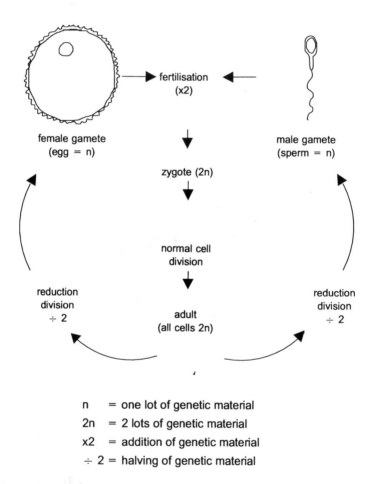

Fig. 2. The sexual cycle in animals. Note that the contribution of the egg to the zygote is much greater than that of the sperm in terms of mass but that the genetic contribution of each is the same.

which can fertilise an egg. In higher plants there are large numbers of pollen grains doing the same job. Which gamete fuses with which is totally unpredictable. Consequently the exact genotype of the zygote is also unpredictable. Any number of genetically distinct zygotes can be produced.

▶ *Recombination during reduction division* – Not only can the genetic material be divided in different ways when gametes are formed but it can also be mixed up so that different parts recombine together.

A combination of these three sources of variation ensures that sexually produced progeny are genetically dissimilar. The almost infinite potential for sexually generated variation has been both a driving force and one of the biggest problems for geneticists.

The germ line

In sexually reproducing multicellular organisms inheritance can only occur through what is known as the germ line. The germ line is the continuity of cells that lead to the production of gametes. Since gametes are the only carriers of genetic information from one generation to the next, they and the cells that they come from, are paramount in determining the genetic makeup of the next generation.

Suppose a mutation (a rare and accidental change in the genetic information) were to occur in a cell that is part of the germ line. In time the mutation would appear in a gamete. If the gamete fused with another to form a zygote, the zygote would develop into a mutant individual. The mutation would have been inherited and fixed in the genetic information of a whole new individual.

What if the same mutation occurred in a cell of the body that was not part of the germ line – say a human skin cell or a cell of the root of a plant? In this case it would exist only as long as the life of the owner of that cell. It could not be inherited by the next generation.

It is because of the uniqeness of the germ line that the idea of inheritance of acquired characteristics is not generally supported. The children of an old man are not born wrinkled or grey haired. Changes to the body of an organism do not affect the germ line. In a sense the germ line is ageless.

Continuity and change in heredity

▶ *Genetics creates a paradox* – It is responsible for both uniformity and variation.

Life on earth today is not very different from what it was a thousand years ago. There are still the same species present. Any changes to the variety of types of living organism are insignificant compared with the overall similarity of life as a whole. And yet there is hardly an individual of any species still living today that was alive a thousand years ago. Most living things will have gone through hundreds or thousands of generations in that time.

This is perhaps the most striking feature of life – its ability to perpetuate itself more or less unchanged for generation after generation. It is a tribute to the efficacy of the genetic mechanism.

Looking closer, however, we see that everything is not the same. No human being today is exactly the same as anyone who lived previously. No two people, with the exception of identical twins, are ever exactly the same. Beneath the uniformity there is variation.

Because genetic studies tend to concentrate on the differences between living things of the same species, for example whether a plant has red flowers or white flowers or whether a bacterium is resistant to an antibiotic or not, the overall similarities are understated. There *are* genetic differences between individuals of one species and genetic mechanisms are constantly adding to these differences, but the degree of similarity is always far greater than any difference. Human beings, supposedly, share about ninety eight per cent of their genetic information with gorillas. This shouldn't surprise us. What should surprise us is that it is not more.

▶ *Key point* – Genetic systems are essentially conservative. They maintain similarity very accurately over long periods and many generations. At the same time they generate small amounts of variation through sexual processes.

The benefits of variation

A small degree of variation in a species gives the species adaptability. This is a big topic which we will touch upon in later chapters. Ultimately genetic variation is the prerequisite for evolution. Without variation there would be no change. But we know that change does happen over time and evolution theory suggests that, if there is enough time, the change can be very great.

The requirements of a genetic material

Before looking at the next chapter, why not pre-empt it by asking yourself what sort of material the carrier of genetic information needs to be? We know, for example, that genetic information survives, without change, through countless replications, but that it can be altered in certain circumstances. It has to carry huge amounts of information but fit neatly inside a cell. It has to be common to all living things. Genetic engineering tells us that genetic material from one organism can fit quite happily into that of any other organism and behave as if it had always been there.

See if you agree with the following list of requirements for a genetic material:

A molecule which:

(a) is common to all living systems
(b) can replicate itself and go on replicating time after time without error
(c) is very large (remember it has to carry a lot of information) but . . .
(d) can be neatly packaged
(e) is infinitely variable (though all its different varieties will share a common basic structure)
(f) contains a code or language which can record information (somehow a plan for all the characteristics of an organism has to be encapsulated in its molecular structure)
(g) is linked to a system for interpreting the information it contains and for putting it into action
(h) is stable but . . .
(i) subject to occasional change

Judge for yourself by reading the next chapter whether DNA meets these requirements for a genetic material.

A brief history of genetics

It is useful to have some idea of how our present knowledge of genetics and its application is based on a chronological sequence of steps in understanding. Some of these steps are small, some seminal, but all lead in the same direction and all are linked like the rungs of a ladder.

It is also worth reflecting that, by the end of the twentieth century, it had become routinely possible to transfer a genetic trait from one organism to another, to identify individuals by means of traces of their genetic material and to document the complete genetic complement of whole organisms. Yet no more than fifty years earlier the identity of the genetic material was still in doubt. The second half of the twentieth century has undoubtedly been a revolutionary period for the advancement of genetics.

The more important advances in genetics are listed chronologically in the table at the end of this chapter (fig. 3).

Tutorial

Progress questions

1. What is the distinction between classical genetics and molecular genetics?

2. What do the terms genotype and phenotype define? How are genotype and phenotype related?

3. What are the important genetic differences between asexual reproduction and sexual reproduction? How does the sexual process produce variation?

4. Why is meiosis also called reduction division?

5. What is meant by the germ line? Why is the germ line so important in genetics?

Seminar discussion

1. Is the idea of the inheritance of acquired characteristics (known as Lamarckism) compatible with a modern knowledge of genetics?

2. Which do you think is more important in determining phenotype: genetics or environment?

3. Living organisms use both asexual and sexual means to reproduce. Why?

Practical assignments

1. Find out what you can about the cloning of animals and the potential for cloning human beings.

2. Choose a plant which is easy to propagate vegetatively and has several distinct varieties. African violet (*Saintpaulia*) or *Pelargonium* would be good choices. Grow new plants from cuttings of phenotypically distinct varieties to see if they show the same characteristics as the parent plant – as you would expect from asexual reproduction.

Study tips

Try to keep a broad view of genetics. Genetics relates to every area of biology and this should be reflected in your study of the subject.

Important advances in genetics

Ancient period
Aristotle recognised that inheritance is the passing of *information* to the embryo.
General recognition that 'like begets like'.
No real experimental genetics.

1850s
Rudolf Virchow's dictum 'every cell comes from another cell'.
Nucleus of cell is seen as seat of cellular heredity.
Sperm and egg are seen as having equal hereditary contributions to make to the next generation.
Charles Darwin's theory of evolution by natural selection (1859).

1860s
Gregor Mendel's paper 'Experiments on plant hybrids' (1866) proposes the particulate nature of inheritance and becomes the foundation of classical genetics.
Friedrich Miescher discovers nucleic acids (1868).

1880s
Chromosomes discovered.
Mitosis and meiosis described.
Wilhelm Roux argues that chromosomes constitute the hereditary material and postulates a linear arrangement of hereditary units along chromosomes.
Egg and sperm contribute an equal number of chromosomes to a zygote.

1890s
August Weismann develops the chromosome theory of heredity.

1900s
Mendel's work is rediscovered and confirmed by both Hugo de Vries and Carl Correns – this marks the real starting point for classical genetics.
Mutation proposed as source of changes in genes (1901).
Archibald Garrod makes the first connection between genes and enzymes and, thereby, initiates the biochemical view of genetics (1902).
Debut of the word 'gene' to describe a hereditary unit (1909).

1910s
Thomas Morgan starts genetic studies using the fruit fly *Drosophila melanogaster*.
The discovery of X Y sex determining mechanism.

1920s
Recognition of two types of nucleic acids DNA and RNA.
X rays cause mutation in fruit flies (1927).

Fig. 3. Table showing important advances in genetics.

1930s
In cells almost all the DNA is in the nucleus, all the RNA in the cytoplasm.
Discovery of linkage and recombination of linked genes by crossing over.
First genetic mapping of a chromosome.
Following the work of Garrod – further investigation of the links between genes and biochemical pathways.

1940s
Birth of bacteriophage genetics leading to isolation of phage mutants and demonstration of recombination in viruses.
Discovery of genetic transformation provides the best evidence yet that DNA is the genetic material.
George Beadle and Edward Tatum come up with the 'one gene – one enzyme' theory.
Proof of spontaneous origin of bacterial mutants – consequent death of Lamarckism.
First genetic cross between bacteria (Joshua Lederberg and Edward Tatum, 1946).
Linus Pauling and colleagues shows that the hereditary disease sickle cell anaemia is due to abnormal haemoglobin, thus strengthening the link between genes and proteins.

1950s onwards
The 1950s and 1960s were a period of very rapid and extensive progress in the field of molecular genetics. The birth of bacterial and phage genetics and the growing understanding of the biochemical relationship between genes and enzymes enabled the bulk of our present knowledge of the structure and function of DNA and RNA to be worked out in the space of less than twenty years. The most significant advances in this time were:

– the structure and replication of DNA
– the mechanism of protein synthesis and the roles of messenger RNA, transfer RNA and ribosomes
– cracking of the whole genetic code
– the molecular mechanism of crossing over
– regulation of gene activity

Since 1970 advances in genetics have been made in the fields of medicine and biotechnology. It is only quite recently that the basic knowledge gained mainly in the fifties and sixties has been married to technologies which enable it to be put to use. It is worth remembering, however, that the basic rules of inheritance have remained unmodified since Mendel formulated them in 1866. The molecular structure of DNA, worked out in 1953, is still accepted today.

Fig. 3. continued.

2

DNA

One minute summary – Central to the study of molecular genetics is the recognition that DNA is the genetic material of practically all cells and organisms. The ability of pure DNA to alter the hereditary characteristics of an organism, and the successes of recombinant DNA technology, leave no remaining doubt as to the genetic role of DNA. All DNA molecules have the same basic structure but vary in the number and precise arrangement of their component parts. The potential variation is so great that an infinite number of different DNA molecules is possible. This means DNA can act as the carrier of genetic information for every possible living thing. In this chapter you will learn about:

▶ DNA – its nature and structure as a genetic material
▶ base pairing as it applies to DNA, and some of its consequences
▶ how DNA replicates
▶ 'semi conservative' replication of DNA
▶ the repair of DNA damage
▶ evidence that DNA is the genetic material

What is DNA?

The nature and structure of DNA as a genetic material

DNA stands for deoxyribonucleic acid. The term nucleic acid was coined in the latter part of the nineteenth century to describe a new class of substance found in living matter, predominantly in the nuclei of cells. Deoxyribo refers to the fact that DNA contains the sugar deoxyribose. Deoxy simply means 'missing oxygen' and signifies that deoxyribose contains one less oxygen atom per molecule than the very similar sugar, ribose. So, DNA is a deoxyribose containing acid found in the nuclei of cells.

Of course, it is a lot more than that. It is the molecule at the centre of all genetics. There may still have been doubt, right up to the middle of the twentieth century as to the importance of DNA, but it is now the unquestioned tenet of genetics that DNA is *the* carrier of genetic information in all living systems other than some viruses.

Variation in DNA

DNA is not really one substance. It is an infinite number of different substances of the same type. The size of DNA molecules can vary and so can the precise arrangement of the components of different molecules. The language of DNA has as great a potential to write molecules as the English language has to write books.

The structure of DNA

All DNA molecules are made up of assemblages of units attached together, end to end, like a string of beads. The resultant molecule is very long and thin. The variation comes in the precise order of different units and in the total number of units making up the molecule. Each unit consists of three subunits. They are:

1. the sugar deoxyribose
2. a phosphate group
3. one of (with a very few exceptions) four nitrogenous bases; the four possible bases are: adenine, guanine, cytosine and thymine.

When these three subunits are joined together they make a nucleotide. Since there are four different types of base, there are four possible types of nucleotide. There is only one form of deoxyribose in DNA and one form of phosphate group, so the only variation comes from the different bases.

Is chemistry necessary?

You will find no chemical formulae in this book. If you wish to know the chemical structure of all the bases of DNA and of the sugar and phosphate parts there are plenty of references to go to. But to understand the genetic role of DNA it is not necessary to know the detailed chemistry of the molecule.

Making a DNA molecule

Fig. 4 shows how, when deoxyribose is joined to phosphate and a base

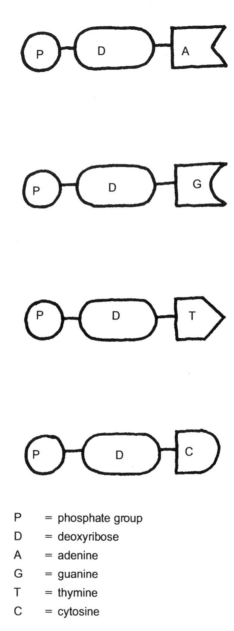

P = phosphate group
D = deoxyribose
A = adenine
G = guanine
T = thymine
C = cytosine

Fig. 4. Four types of nucleotide making up DNA.

one of four types of nucleotide is formed. All we have to do to make a DNA molecule is to join lots of nucleotides together. In joining them together it is the sugar part of one nucleotide that connects up to the phosphate of the next.

Sizes of DNA molecules

In fig. 5 the four possible types of nucleotide are joined together to form one very small DNA molecule. To give an idea of how small this is, compared to other DNA molecules: the DNA from one type of bacterial virus has a length of approximately two hundred thousand nucleotides. Even this is not a particularly large DNA molecule. DNA from a bacterium can be a single molecule four million nucleotides long.

Arrangement of nucleotides

The nucleotides that join together to make a DNA molecule can do so in any order. There are 24 different ways in which fig. 5 could have been drawn. You can work these out for yourself, bearing in mind that AGTC, for example, is not the same as CTGA, because the chain of nucleotides has a definite direction to it.

A four nucleotide piece of DNA, however, does not have to contain four different nucleotides. It could be AAAA or AATT or CCCG or any other combination of A, T, G or C.

If a piece of DNA with just four units can show so much variety, the variation possible in chains of hundreds of thousands of nucleotides is almost unimaginable. This is what gives DNA its potential to convey huge amounts of information.

Base pairing

So far we have considered DNA as a single chain of nucleotide units. DNA does sometimes exist naturally in this form. Some viruses, for example, have single stranded DNA. More usually, however, DNA is in the form of two strands running side by side (see fig. 6). The two strands run in opposite directions and are held together by hydrogen bonding between bases. Hydrogen bonds are the same forces that hold water molecules together and give water its cohesive properties.

Hydrogen bonds are not as strong or permanent as the chemical bonds that keep the nucleotides making up each strand of DNA

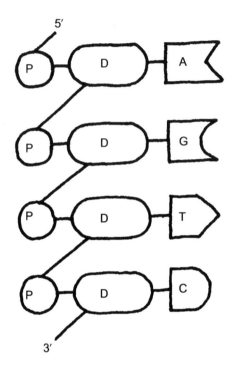

Note:

(i)   The connection between the phosphate group and the deoxyribose of the adjacent nucleotide.

(ii) The potential for attachment of other nucleotides at each end.

(iii) 3′ and 5′ are notations signifying the direction of the nucleotide chain. The numbers refer to the third and fifth carbon atoms of the deoxyribose molecule. You will see that the 5 carbon in this chain of nucleotides is at the top and the 3 carbon at the bottom. The DNA molecule, therefore, runs from the 5′ to the 3′ end down the page.

Fig. 5. Nucleotides joined in line to make a DNA molecule.

together, but are sufficiently strong to hold the two partner strands in place like the two sides of a zip fastening. Because they are not proper chemical bonds they can be broken by physical means such as heating. When a solution of double stranded DNA is heated above a certain temperature it is said to denature or 'melt'. In other words the paired strands fall apart and the DNA becomes single stranded. The DNA from different sources has different melting temperatures but, typically, melting happens at about 70° C.

Specificity of base pairing

In double stranded DNA the pairing of bases is normally specific. Adenine (A) only pairs with thymine (T) and vice versa, while guanine (G) only pairs with cytosine (C) and vice versa.

The rule of specific base pairing is crucial to the function of DNA. Wherever there is the base adenine in one strand of double stranded DNA there must be the base thymine in the other and vice versa. Wherever there is guanine there must be cytosine opposite it and vice versa. If the base sequence of one strand is known the base sequence of the other can be worked out. The two strands are said to be complementary.

DNA hybridisation

When DNA is denatured by heating (turned from double stranded to single stranded) and then cooled, the single strands rejoin as they were before. This is called renaturing. It happens because the strands are complementary and base pairing is at its maximum when they are paired in their original form.

Single strands of DNA from different sources will pair together only if they are complementary or complementary over much of their length. They form what is known as hybrid DNA. The ability of different types of DNA to hybridise can be used as a test of their similarity. The use of gene probes depends on the ability of complementary DNA base sequences to recognise each other and bind together (see chapter 10).

Some other facts about DNA

Circular DNA

At one end of a DNA strand there is a vacancy for attachment of a phosphate group and at the other end there is a vacancy for attachment of a deoxyribose sugar. So a strand can form a

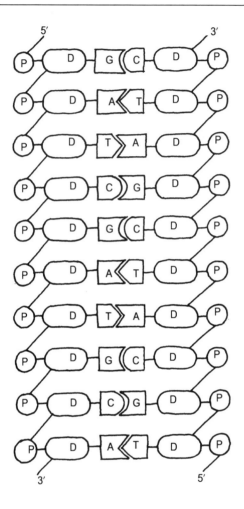

Note:

(i)   The two strands are held together by pairing of bases.

(ii)   The force of pairing comes from hydrogen bonds.

(iii) The two strands run in opposite directions.

(iv)   The sequence of bases along one of the strands has been chosen at random, but . . .

(v)   Once the sequence of bases in one strand is specified there can only be one possible sequence for the other strand. The reason for this is the rule of specific base pairing.

Fig. 6. A piece of double stranded DNA.

connection end to end and the result is a complete circle of DNA. Circular DNA is found in bacteria, for example.

Base equivalence rule

If the molar proportion of one of the DNA bases in a piece of DNA is known the molar proportions of the other three can be worked out. For example, if adenine molecules make up 22% of the bases in a DNA molecule, then the specific base pairing rule means that thymine must also be 22%. That leaves 56% guanine and cytosine combined, and since these must also be equal they comprise 28% each.

The observation that DNA always contains roughly equal numbers of the bases adenine and thymine and also equal numbers of guanine and cytosine was the clue that first led to the idea of specific base pairing:

$$A = T \text{ and } G = C.$$

Hydrogen bonding between bases

When bases pair in double stranded DNA, there are three hydrogen bonds between G and C but only two between A and T. In other words the attraction between G and C is stronger. This is evident in the different melting temperatures of different types of DNA. The higher the proportion of G – C pairs in the molecule the higher the melting temperature.

Distance between adjacent nucleotides

The actual distance between adjacent nucleotide pairs along the length of a DNA molecule is 3.4 Angstrom units, or 0.34 millionths of a millimetre. So a millimetre length of DNA would contain approximately three million nucleotide pairs. In a single human cell there is estimated to be a total of between one and two metres of DNA! You can work out how many nucleotides that is – but bear in mind that in a human cell there are usually two copies of every piece of its DNA: so divide your answer by two to get the number of nucleotides in a single set of human genetic material.

The double helix

The two strands of a DNA molecule are believed to coil around each other in a spiral manner, with ten nucleotide pairs per single turn of

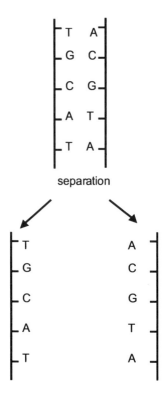

separation

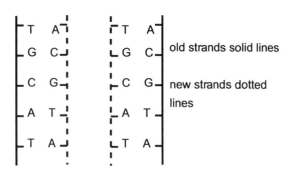

synthesis of new strands

old strands solid lines

new strands dotted lines

Fig. 7. Summary of replication of DNA. Note that each new strand is an exact copy of one of the original strands.

the spiral. Evidence for this configuration comes largely from X-ray diffraction pictures. None of the properties of DNA that are relevant to its genetic role are dependent on this idea. In fact it seems to complicate the operation of DNA activity quite considerably, without conferring any obvious advantage.

For simplicity's sake, all the diagrams of DNA given here show the strands of DNA as untwisted structures. This should not detract from their information content in any significant way.

How DNA replicates

The genetic material must be able to make exact copies of itself. The rule of specific base pairing in double stranded DNA is the key to the molecule being able to do this. DNA replication can be summarised in one sentence: 'The two strands of the molecule separate and a new partner is made for each one, following the rule of specific base pairing' (fig. 7).

The process of replication, though very simple in summary, is actually quite complicated. It has been estimated, for example, that at least fifteen different proteins participate directly in DNA replication.

The need for accuracy in DNA replication

DNA replication has to be accurate. The replication system includes complex devices to eliminate the possibility of mistakes, or to correct them if they happen. The actual error rate is calculated to be of the order of one per ten billion base pairs copied.

Replication is as accurate as this for two reasons. Firstly, it is template directed. That is, the old strands of DNA are a template, or instructions to be followed, for making the new strands. Secondly, errors are recognised and corrected as replication proceeds.

The replication process in more detail

The simple summary we began with contains all that is vital to a genetic understanding of DNA replication. We can fill in at least some of the details with the following account (and see fig. 8).

1. *Conditions for replication*
For DNA replication to take place there must be:

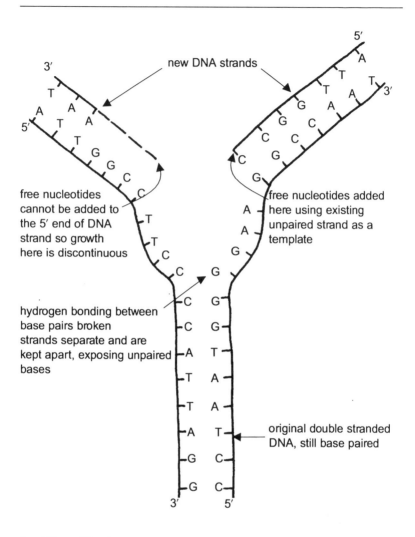

free nucleotides
cannot be added to
the 5′ end of DNA
strand so growth
here is discontinuous

free nucleotides added
here using existing
unpaired strand as a
template

hydrogen bonding between
base pairs broken
strands separate and are
kept apart, exposing unpaired
bases

original double stranded
DNA, still base paired

Q　What will be the next nucleotide added to the new DNA strand on the right
of the diagram?

A　It will be a nucleotide with the base thymine (T), since the next unpaired
base on the original DNA strand is adenine (A). Note that we cannot ask
the same question with reference to the left hand strand because that
one is not growing step by step in the same direction as the replication
fork is opening up. It is growing discontinuously in the opposite direction.

Fig. 8. DNA replication: events at the replication fork.

(a) a pre-existing DNA template

(b) all four types of nucleotide in an activated state (activated, in this case, means having two extra phosphate groups attached to them – a source of energy for the reaction)

(c) the necessary enzymes. When DNA was first synthesised in a cell free (in vitro) system only one enzyme, DNA polymerase 1 was added. Within cells (in vivo) it is believed that more enzymes are involved.

2. *Separation of strands*
The two strands of the DNA that is about to be replicated separate from each other. In the circular DNA of bacteria such as E. coli this separation starts at one specific region. In the DNA of higher organisms the strands may separate in several places simultaneously, so that replication starts in more than one place at the same time. More than one protein is involved in the job of breaking the hydrogen bonds between bases, opening the double stranded DNA and keeping the two strands apart as synthesis of new strands proceeds.

3. *Synthesis of new strands*
When the strands of DNA are separated the bases no longer pair with each other in that region. So there is potential for each unpaired base to pair with a new base of the right kind (A with T, G with C). Suitable bases are present as free activated nucleotides. All that is required is for these nucleotides to be slotted into place and combined together in two brand new DNA strands. A DNA polymerase enzyme and other proteins do this job.

4. *Direction of synthesis*
DNA synthesis can proceed in both directions away from the starting point provided the starting point is not the end of the DNA. In a circular piece of DNA, of course, there is no end. Where replication is taking place the original double stranded DNA is split down the middle and the two arms of newly synthesised DNA spread out in a Y shape. This branching point is referred to as the replication fork.

5. *Distinction between strands*
The DNA polymerase enzyme can only add nucleotides to a growing

DNA strand in one direction. But the two newly synthesised strands are growing in different directions. To get round this problem one strand grows continuously while the other grows discontinuously.

6. *Dealing with errors*
DNA polymerase can not only catalyse the synthesis of DNA strands but also their breakdown. This enables the enzyme to undo and correct any errors in replication.

Conclusion
At the completion of replication there is nothing holding the two new double stranded DNA molecules together. They are independent replicas of the original.

'Semi conservative' replication of DNA

A feature of the replication of DNA, which should now be obvious, is that when two molecules are produced from a single one, each new molecule actually contains half of the original. This method of replication is referred to as semi conservative. If, on the other hand, replication were conservative the original molecule would be unchanged. A totally new molecule would be formed of which no part came from the original (see fig. 9).

Semi conservative and conservative replication of DNA
The Meselson-Stahl experiment
If new and original DNA could be told apart in some way, it would be possible to test which type of replication, conservative or semi conservative, is used by living systems.

Exactly this was done in 1957 in a famous experiment known, after the people who did it, as the Meselson-Stahl experiment. Meselson and Stahl grew bacteria for several generations in the presence of nitrogen compounds containing the isotope of nitrogen, ^{15}N, in place of the normal ^{14}N. This ensured that all the DNA in the bacteria was made with ^{15}N nitrogen. This is the original DNA. They then transferred the bacteria to a medium containing only normal ^{14}N nitrogen.

As the bacteria continued to synthesise DNA the new strands were made with ^{14}N nitrogen. DNA containing ^{15}N is noticeably denser

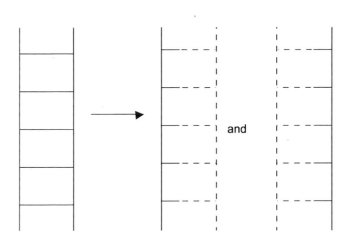

1. semiconservative

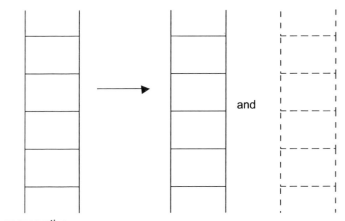

2. conservative

Fig. 9. Semi-conservative and conservative replication of DNA. Old strands solid lines, new strands dotted lines.

Relative densities and amounts of DNA extracted from bacteria at start of experiment and at different times thereafter.

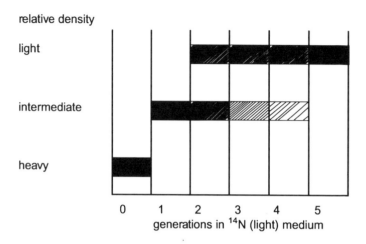

Fig. 10. Result of the Meselson-Stahl experiment.
The degree of shading shows relative amount of DNA.

than ^{14}N DNA. The former is 'heavy' DNA and the latter, 'light' DNA. Nitrogen, ^{15}N, contains an extra neutron in each atom compared with ^{14}N nitrogen. It's mass is, therefore approximately one fourteenth as great again, and any molecule containing it will weigh more.

Following the transfer of bacteria from a medium with heavy nitrogen (^{15}N) to a light nitrogen (^{14}N) medium, Meselson and Stahl took samples at various times, extracted the DNA from the bacteria and measured its density.

The crucial finding was that after exactly one generation of growth in the new medium – when every piece of DNA had replicated precisely once – the extracted DNA was intermediate in density between heavy DNA and light DNA. In other words it contained heavy and light DNA in equal proportions.

The most likely explanation is that, of the two strands of the one generation DNA, one is from the original, heavy DNA and the other is newly synthesised light DNA. This is what is expected if DNA replication is semi conservative.

► *Key point* – The Meselson-Stahl experiment confirmed the semi conservative model of DNA replication, in which each parental strand of the DNA ends up in one of the two copies, paired with a newly synthesised, complementary strand.

The repair of DNA damage

It is vital for the survival of an organism that the integrity of its genetic material is maintained. Changes can and do happen to DNA, almost all of which are more likely to be harmful than beneficial. Occasionally there is an error in replication. Or damage can be done by chemical substances or physical influences. To preserve the DNA in its proper form a number of repair mechanisms operate, constantly checking the DNA, looking for damage, and then repairing it.

Ultra violet light damage and repair

Ultra violet light is known to kill bacteria and to do so by damaging the DNA. Ultra violet lights are often used to keep surfaces and materials sterile. The most effective wavelength of light for killing bacteria is also the one most strongly absorbed by DNA.

The effect of UV light on DNA is to cause adjacent thymine bases on the DNA strand to combine together to form a thymine dimer. The presence of thymine dimers renders the DNA ineffective. But bacteria exposed to damaging amounts of UV light can recover their function due to repair mechanisms. An enzyme checks the DNA, locates the presence of thymine dimers and cuts them out. Types of E. coli that lack this enzyme cannot repair UV damage and so are UV sensitive. The gap made by excision of a thymine dimer is then patched by enzymes that mimic the replication process, repositioning normal thymine nucleotides using the complementary DNA strand as a template.

This 'cut and patch' method of repair – removing the damaged DNA and replacing it with a new, undamaged piece – is common to most types of damage that DNA is subjected to.

Two strands provide insurance

The double stranded nature of DNA ensures that if one strand is damaged there is still enough information in the other strand to repair the damaged one. This is exactly the same principle used in replication. In fact, the complementary nature of the two DNA strands is, itself, a safeguard against permanent change to the genetic material. In the same way that important documents are always duplicated or computer material kept on both hard and floppy discs, DNA keeps two copies of its information, one in each strand.

▶ *Key point* – Damage to one strand of DNA does not mean the irretrievable loss of the genetic information because a copy is kept in the complementary strand.

Evidence that DNA is the genetic material

The success of modern DNA technology is evidence enough that DNA carries genetic information. Indeed, it may seem a completely redundant exercise to go over the background to the concept of DNA being the material of inheritance. The role of DNA is now simply taken for granted.

Nevertheless, it is useful to view the study of DNA from a historical perspective. And exam questions still sometimes ask for the *evidence* that DNA is the genetic material!

Circumstantial evidence that DNA is the genetic material

1. *DNA is in the nucleus*

DNA is located primarily in the nuclei of cells. Indeed, the discovery of nucleic acids came about because it was realised that the nucleus was all important in heredity and its contents were investigated for that very reason.

2. *DNA is in chromosomes*

Within the nuclei of cells, DNA is specifically located in chromosomes. The behaviour of chromosomes at mitosis and meiosis exactly parallels the behaviour of the hereditary units. Chromosomal changes and abnormalities are often seen to be linked to inherited characters.

3. *Constant amount of DNA per chromosome set*

The amount of DNA in a cell is proportional to the number of sets of chromosomes in that cell. So, for example, a sperm or egg cell contains half the amount of DNA found in a zygote (see fig. 1).

4. *DNA in cell organelles*

Some cell organelles contain small amounts of DNA. Mitochondria and chloroplasts are examples. These organelles can show inheritance of characters which is independent of nuclear inheritance.

5. *Mutation*

Substances or influences that cause genetic mutation also cause chemical changes to DNA. Ultra violet light is a good example (see DNA Repair, p 41).

6. *Structure of DNA*

The structure of DNA satisfies all the requirements of a genetic material.

7. *Replication*

DNA can replicate itself. No other component of cells can do this.

Direct evidence that DNA is the genetic material

1. *DNA technology*

The best direct evidence that DNA is the genetic material comes from recent advances in DNA technology. These will be covered in more detail later in the book. In recombinant DNA technology and in the production of transgenic organisms, in essence, what is done is to transfer DNA from one organism into another and, thereby, donate some of the characters of the donor organism to the recipient. The facts that the functions of an organism can be altered by pure DNA and that DNA, in whatever organism it finds itself, retains its original functional role is evidence enough that the hereditary characters of living things are incorporated wholly and solely in DNA.

2. *Genetic transformation*

The ability of pure DNA to 'transform' the genetic type of an organism was first discovered in the 1940s. In the period leading up to the unchallenged acceptance of DNA as the genetic material, the study of genetic transformation provided the first and, at the time, the best direct evidence for the role of DNA.

The story starts with the work of Frederick Griffith, begun in 1927. He experimented on two types of the bacterium that causes pneumonia, called 'pneumococcus' S type and 'pneumococcus' R type. S type bacteria produce smooth, shiny colonies in culture because they make a thick, gelatinous capsule around their cells. R type bacteria, on the other hand, have rough colonies because they do not make such a capsule. The capsules of S type bacteria protect them from the body's immune system. For this reason S bacteria are extremely pathogenic, whereas R bacteria are harmless.

Mice injected with the virulent S type pneumococcus all died within two days. Mice injected with the R type remained healthy. When Griffith injected mice with a mixture of living R type bacteria (harmless) and dead S type (also harmless because dead bacteria cannot cause disease) the mice died. What made this part of the experiment even more interesting was that from the dead mice he was able to extract living bacteria of both R and S types. Either the dead S type bacteria had come back to life (unlikely) or else the living R types had been 'transformed' into S types. Griffith's experiment and the results he obtained are summarised in fig. 11.

▶ *Key point* – Dead bacteria can transform living ones. The characters passed from dead to living are permanently incorporated and inherited by future generations.

The component of dead bacteria responsible for transformation became known as the 'transforming principle'. Its identity would lead directly to the genetic material, as only a genetic material could pass on information that both transforms the immediate recipient, and does so in such a way that the change is inherited.

Those who followed Griffith ultimately identified the transforming principle as DNA. Three co-workers, Oswald T. Avery, C. M. McCleod and M. J. McCarty, adopted the logical approach of separating and purifying the various components of dead bacteria that might be active and testing them one by one. When they removed all materials other than DNA from extracts of dead bacteria the transforming principle remained intact. When they removed DNA there was no transformation.

It was found that as little as one part in six hundred million of the purified DNA from one type of bacterium could effect transformation and that the DNA could be purified until it contained a mere 0.02% protein without losing any of its activity.

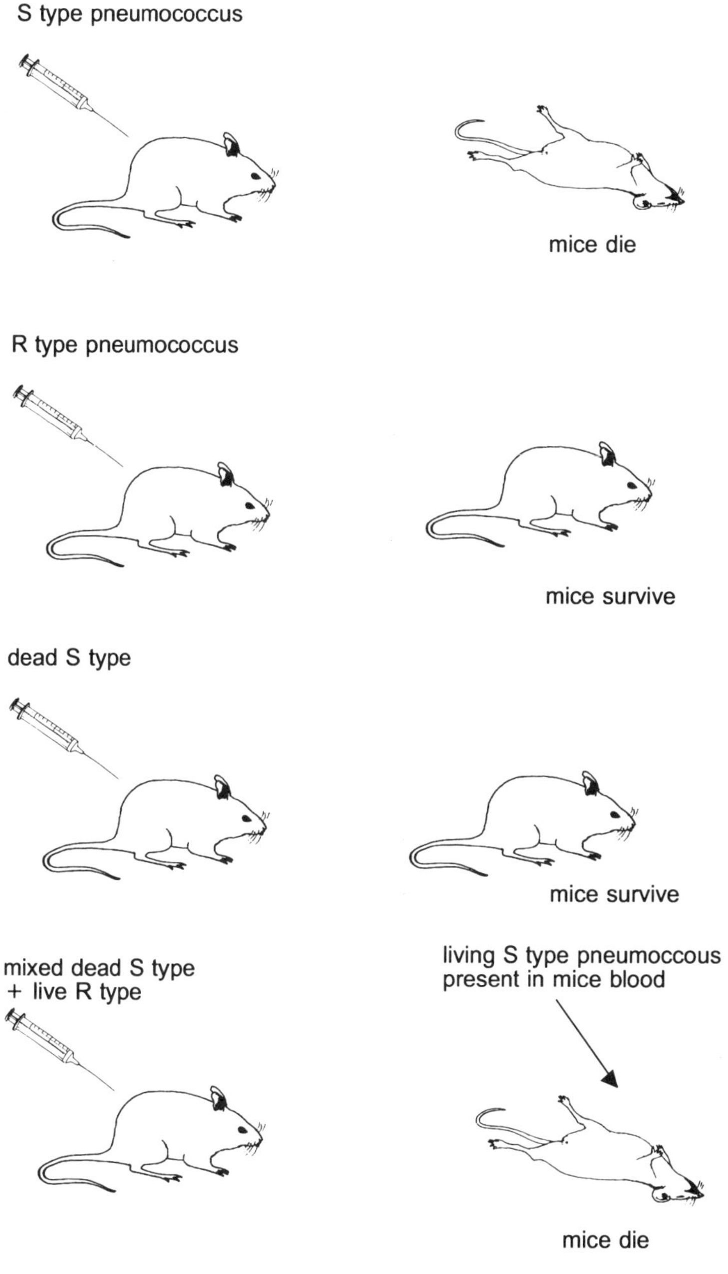

S type pneumococcus

R type pneumococcus

dead S type

mixed dead S type
+ live R type

mice die

mice survive

mice survive

living S type pneumoccous
present in mice blood

mice die

Fig. 11. Griffith's experiment: the discovery of transformation.

3. *The Hershey-Chase experiment* – To understand the famous experiment of Alfred Hershey and Martha Chase it is necessary to know something about bacterial viruses, or bacteriophages as they are known (phages for short). It may, therefore, be helpful to read this account in conjunction with the chapter on viruses (Ch. 4). The bacteriophage used by Hershey and Chase is approximately half protein and half DNA. It infects the bacterium E. coli and, after a period of time following infection, the host bacterium breaks apart, liberating about two hundred new phage particles. Clearly the instructions that are followed in the manufacture of these new phages must enter the body of the host bacterium, since this is where the new phages are made. These instructions constitute the genetic information. Perhaps the whole virus enters the bacterial cell, or just a part of it. If only a part of it then perhaps just the DNA or just the protein. Hershey and Chase set out to determine which part of the virus enters the bacterial cell and, therefore which part carries the genetic information. A detailed scheme for the Hershey-Chase experiment is shown in fig. 12. But first a summary of the conclusion: *When a bacteriophage infects a suitable bacterial host, the DNA of the phage enters the bacterial cell and the protein component of the phage remains behind. It is, therefore, the DNA that carries the information needed for the manufacture of new phages.*

The key points to understanding the Hershey-Chase experiment are:

(a) DNA contains phosphorous but protein doesn't.
(b) Protein contains sulphur but DNA doesn't.
(c) Phages were labelled with either ^{35}S or ^{32}P, not both. There were, effectively, therefore, two separate experiments, one in which the protein was labelled and one in which the DNA was labelled.
(d) The purpose of blending was to separate phage particles from the bacteria to which they had attached themselves, but not before some part of them had entered the host and caused infection.
(e) The purpose of centrifugation was to isolate the part of the phages that had entered the bacteria from the part that remained outside.
(f) It was vital to show that the sedimented bacteria really were infected and could produce new phage particles.
(g) The shearing force of the blender treatment stripped 75 to 80% of the 35S labelled protein from the surface of the bacteria. That leaves 20% unaccounted for. The best conclusion from the evidence is that most of the phage DNA enters the host and that most of the protein does not.

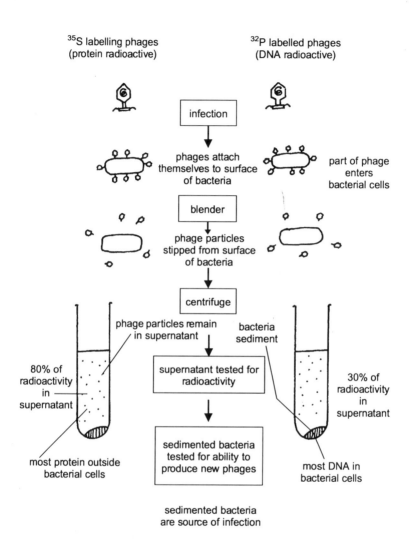

Fig. 12. The Hershey-Chase experiment.

4. Transfection

Subsequent to the Hershey – Chase experiment it has been found that purified viral DNA, by itself, can, in certain circumstances, cause infection of bacteria and the production of new phage particles. The process is known as transfection and proves conclusively that DNA alone carries all the information needed to specify the whole viral structure.

The universality of DNA?

Strictly, transformation experiments only showed that DNA is the genetic material of bacteria and the Hershey – Chase experiment and transfection show that the same is true for some bacteriophages. It is only as a result of more recent work, such as the technique of genetic engineering, that clear direct evidence has emerged that DNA is the universal genetic material.

Tutorial

Progress questions

1. Show by means of a diagram how deoxyribose units, phosphate groups and four types of base fit together to make a double stranded DNA molecule. Can you identify a single nucleotide on your diagram?

2. If the base sequence in one strand of a double stranded DNA molecule is

    ATTGACGGAACTTGC

    What is the base sequence on the other strand? Are these sequences going in the same direction? What word is used to describe the relationship between the base sequences on two paired strands of DNA? The two strands are . . .?

3. DNA replication is semi conservative. What does this mean? Explain how the rules of specific base pairing enable DNA to replicate exactly.

Seminar discussion

1. What would you consider the best evidence that DNA is the genetic material? How good is the rest of the evidence?

2. Look back to the end of the first chapter. How well does DNA fit the criteria for a genetic material?

Practical assignment

Two experiments described in this chapter use isotopes to label molecules. The use of isotopic labelling is common in biological research. Make sure you understand (a) what an isotope is, and (b) how the presence of isotopes in molecules can be detected.

Study tips

1. To remember which bases pair with which in DNA, think of straight lines and curves. A and T are straight line letters; C and G are curved.

2. Don't confuse thymine, the DNA base, with thiamin, the B vitamin; nor make the mistake of thinking the DNA bases are amino acids just because their names sound similar.

3. Diagrams and models of DNA can look very complicated. But, if you understand the importance of base sequence and specific base pairing, the essential components are extremely simple. You can reduce your diagrams to a pair of lines to represent the sugar-phosphate backbones of the DNA strands and a sequence of letters to show the bases, without losing any information value (see the diagrams in this chapter).

3

Genes and Proteins

One minute summary – DNA contains a genetic plan. The plan is put into action when the information in DNA is used to make proteins. Proteins, particularly enzymes, control the activities and, ultimately, all the characteristics of cells and organisms. One definition of a gene is that part of a DNA molecule that carries the information specifying a single protein. Built into the molecular structure of DNA is a genetic code in which each word of the code translates into one of the amino acids that make up a protein. The DNA code can, therefore, be read as a sequence of amino acids. When a gene is switched on, the genetic code is read, the encoded amino acids are joined together in the correct sequence and the prescribed protein is made. In this chapter you will discover:

► how the genetic plan in DNA is implemented
► that the information in DNA is used to make proteins
► the importance of enzymes
► the structure of proteins
► the twenty amino acids that occur naturally in proteins
► the nature of the genetic code
► how proteins are synthesised
► one way to define a gene
► how gene function is controlled

How does DNA work?

The first principle of molecular genetics is that the genetic plan for an organism is in its DNA. The second is that DNA reveals its plan through the production of proteins. The information in DNA takes the form of instructions that determine exactly what proteins and, in particular, what enzymes are made by an organism. Enzymes, in turn, control an organism's activities, how it develops and what characteristics it shows.

Biological control

The idea that DNA controls enzymes and enzymes control phenotype is the central dogma of the genetic mechanism. Everything about an organism that has a genetic basis can be explained in terms of the presence or absence of specific enzymes or other proteins. This idea is easy to comprehend in the case of simple biochemical characters, for example, whether or not a bacterium can break down a particular sugar to use as a food source. It is more difficult to imagine how a person's facial features, or a hereditary behaviour pattern, can be explained in terms of enzymes. But the only molecular interpretation of genetic inheritance lies in the concept of DNA directing the synthesis of proteins and enzymes. To make the link between molecular genetics and classical genetics, it has to be assumed that protein synthesis is the basic control mechanism for all inherited characters.

The importance of enzymes

Here is all you need know about enzymes for an understanding of genetics:

1. Enzymes are proteins.

2. Enzymes catalyse biological reactions.

3. Enzymes are highly specific – generally speaking every biochemical reaction that happens in a biological system is controlled by its own particular enzyme.

4. Enzymes' control of biochemical reactions is such that, if the right enzyme is present and all other conditions are suitable, the reaction will proceed; if the enzyme is either not present or non-functional, the reaction will not proceed.

5. Genetic characters can be explained in terms of the presence or absence of functioning enzymes.

The structure of proteins

All proteins have the same basic structure. They are made of lots of amino acids joined end to end in a long chain, like a string of beads. Amino acids are molecules of various shapes and sizes that share common properties. They range in molecular weight from 75 to nearly 190, making them all about as big as, or smaller than, a glucose molecule.

Amino acids

There are twenty different kinds of amino acid found in natural proteins. You will find it useful to memorise this list:

The twenty amino acids commonly found in proteins
(with their three letter abbreviations)

Alanine	(Ala)	Leucine	(Leu)
Arginine	(Arg)	Lysine	(Lys)
Asparagine	(Asn)	Methionine	(Met)
Aspartic acid	(Asp)	Phenylalanine	(Phe)
Cysteine	(Cys)	Proline	(Pro)
Glutamic acid	(Glu)	Serine	(Ser)
Glutamine	(Gln)	Threonine	(Thr)
Glycine	(Gly)	Tryptophan	(Trp)
Histidine	(His)	Tyrosine	(Tyr)
Isoleucine	(Ile)	Valine	(Val)

At one end of an amino acid is an amino $(NH2)$ group and at the other is a carboxyl $(COOH)$ group. The amino group of one amino acid can combine with the carboxyl group of another so that the two amino acids join together. The bond that joins them together is called a peptide bond. As with the structure of DNA, this chapter will deliberately avoid the use of chemical formulae, but the formation of a peptide bond in the combining of two amino acids can be visualised as in Fig.13.

Any of the twenty types of amino acid can join to any other, or to another like itself, by means of a peptide bond. Then, any number of

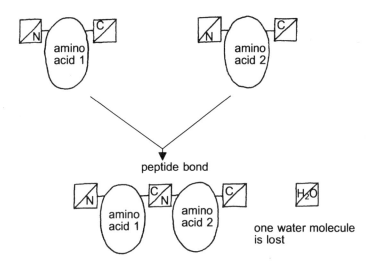

peptide bond

one water molecule is lost

C = carboxyl group
N = amino group

Note:
(i) The bond between two amino acids is formed by combination of the carboxyl group of one and the amino group of the other.

(ii) This still leaves an uncombined carboxyl group on one amino acid and an uncombined amino group on the other, giving the potential for combination with more amino acids. Ultimately a long chain can form.

Fig. 13. Formation of a peptide bond between two amino acids.

amino acids can join in a chain to make a protein. Proteins vary a lot in size but usually contain large numbers of amino acids. Insulin, for example, is made up of 51 amino acids of seventeen different types (fig. 14). The tryptophan synthetase A protein from E. coli , which catalyses one of the steps in the biosynthesis of the amino acid tryptophan, is 268 amino acids long and contains all the different types of amino acids except, interestingly enough, tryptophan itself.

Relationship between DNA and proteins

Both DNA and proteins are linear molecules of great length, made up of units joined together end to end to form a chain. Each individual piece of DNA and each unique protein has a distinct sequence of units, but there are numberless other possible sequences, leading, potentially, to an infinite variety of different DNA molecules and different proteins.

Given that DNA is the first step in the realisation of a genetic plan and that enzymes and proteins control the putting of that plan into action, wouldn't it make sense for the sequence of base pairs in DNA to determine the sequence of amino acids in a protein? This idea is the basis of the way in which DNA exerts its control, and leads to the concept of a genetic code (a code in which the information in DNA relates directly to the amino acid structure of proteins).

One definition of a gene

In molecular genetics a gene is a part of a DNA molecule that carries the information specifying one protein. The gene can be written down as a sequence of nucleotide bases or base pairs. The sequence is the code that leads to the production of one particular sequence of amino acids, characteristic of the protein in question. The gene has a calculable length because we know how far apart the base pairs in DNA are. It is both a structural and a functional unit.

The gene as a controller of biochemical activity

The idea that genes work at a biochemical level developed over a period of a number of years. Archibald Garrod, in 1902, first suggested that a gene might control an enzymatically catalysed metabolic reaction. It later became apparent that:

Fig. 14. The primary structure of insulin. Note that there are two peptide chains in an insulin molecule, the A and the B.

1. Many inherited characteristics of organisms can be explained on the basis of metabolic reactions and their products.

2. Mutations of genes can cause enzymes to cease to function which, in turn, causes a breakdown in metabolism.

3. A gene controls a metabolic reaction by determining the presence or absence of the correct functional enzyme.

4. Because biochemical pathways often have many single steps, they are controlled by many enzymes and, therefore, by many genes.

5. If just one gene controlling one of the steps in a biochemical pathway mutates so as to fail to produce a functional enzyme, the whole pathway is blocked.

6. It follows, therefore, that mutations of different genes can, independently, have the same observable effect on phenotype.

The biosynthetic pathway leading to tryptophan

The amino acid tryptophan is synthesised by the bacterium, E. coli, according to the biosynthetic pathway shown (fig. 15).

Starting with the precursor, chorismic acid, there are six enzymatically controlled steps in the production of tryptophan. This means six different enzymes and six controlling genes.

If any of these genes mutates, thereby failing to make a functioning enzyme, no tryptophan can be produced. The bacterium in question would be a tryptophan requiring mutant (a tryptophan auxotroph) unable to survive without the provision of tryptophan in its food.

Pathway analysis

A mutant of the first gene in the tryptophan pathway (trpA) will be able to make tryptophan if any of the other intermediates (for example, anthranilic acid) are made available to it. A mutant of gene trpB cannot make tryptophan from anthranilic acid but can make it from any of the other intermediates (PRA, CDRP, etc). A mutant of gene trpE must have tryptophan. No other substance will suffice.

Feeding experiments involving different tryptophan requiring mutants of E. coli and different intermediate substances from the

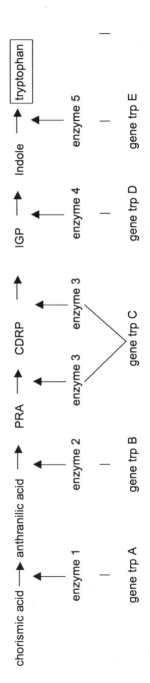

Fig. 15. An example of a gene controlled biochemical pathway: the biosynthetic pathway leading to tryptophan.

biosynthetic pathway can establish both the sequence of the various intermediate compounds, and which step in the pathway is blocked in the case of each mutant type. This is the technique of pathway analysis.

Whenever a single gene mutation blocks a metabolic pathway, the pathway can be restarted by supplying any intermediate that is further along the pathway than the blockage (fig. 16).

The 'one gene one enzyme' theory

In 1941 G. Beadle and E. L. Tatum proposed the idea that each gene has only one function, which is to direct the synthesis of one – and only one – enzyme, and thereby control the single chemical reaction catalysed by that enzyme. This theory became known as the 'one gene one enzyme' theory. It should, perhaps be modified to one gene one protein, since some – albeit few – gene products such as haemoglobin are not enzymes.

The genetic code

The coded message of a gene lies in the sequence of bases or base pairs that make up the DNA. When the coded message is deciphered it translates into a sequence of amino acids that join together to make a protein. Knowing the bases in a length of DNA, you can, therefore, by using the code, work out the primary structure of the protein made by that bit of DNA.

The rules of the code

The genetic code is an example of the most basic type of code, in which each separate code word of the coded message translates directly into one particular meaning. Each code word is a sequence of bases and the translation, in each case, is an amino acid.

An equivalent code would be one in which a number stands for each letter in the alphabet: A = 1, B = 2, C = 3, etc.

Using this code, the 'word':   7   5   14   5   20   9   3   19
would read: G   E   N   ET   I C   S

The genetic code is equally simple. There are, however, rules, even to a code as straightforward as this.

(i)   All genes operational.

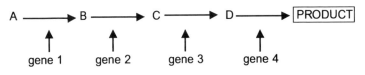

(ii)   Mutation of gene 2 blocks conversion of B to C.

(iii) Pathway restored by addition of substance C or substance D.

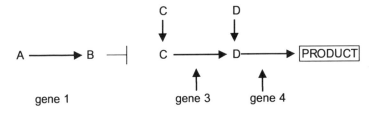

Fig. 16. Restarting a metabolic pathway blocked by a mutant gene. The diagram shows how a mutation of a single gene blocks a metabolic pathway, and how the addition of an intermediate can restore the pathway.

1. It must be read in sequence (7 first, then 5, then 14 and so on).

2. It must be read in one direction only (sciteneg makes no sense).

3. There must be no confusion as to where one code symbol ends and another one starts. 14, for example, translates into n, and must not be misinterpreted as 1 and 4, meaning a d.

4. It must be punctuated (7 is the first letter of one word and 19 is the last; anything after the 19 is the start of a new word, anything before it is part of another, different word).

5. It must not change. Each number must always mean the same letter.

The nature of the code

All the above rules apply to the genetic code. It also has the following features (read this in conjunction with an examination of the code table itself (fig. 17)):

(a) It is a triplet code. This means that a sequence of three bases in DNA specifies each amino acid in a protein. For example the sequence AAA translates into the amino acid phenylalanine, the sequence GAC into leucine, TAC into methionine.

(b) It is non-overlapping. In other words, each base forms part of only one triplet. AAAGAC means phenylalanine (AAA) followed by leucine (GAC) and cannot be read as, for example, AAA (phenylalanine), followed by AAG (which would be phenylalanine again), followed by AGA (serine), then GAC (leucine).

(c) It contains many synonyms. There are 64 possible code words, using the four bases of DNA, three at a time. But there are only twenty amino acids. Either, therefore, some combinations of three bases are not used as code words or, alternatively, more than one code word can signify the same amino acid. The second of these options is the case. So, for example, GGA, GGG, GGT and GGC all translate as proline. There are six code words for leucine. The same is true for arginine and serine. Of all the amino acids, only two, methionine and tryptophan, have a single DNA code word.

first base	second base →				third base
	A	**G**	**T**	**C**	
A	AAA phe	AGA ser	ATA tyr	ACA cys	A
	AAG phe	AGG ser	ATG tyr	ACG cys	G
	AAT leu	AGT ser	ATT *term*	ACT *term*	T
	AAC leu	AGC ser	ATC *term*	ACC trp	C
G	GAA leu	GGA pro	GTA his	GCA arg	A
	GAG leu	GGG pro	GTG his	GCG arg	G
	GAT leu	GGT pro	GTT gln	GCT arg	T
	GAC leu	GGC pro	GTC gln	GCC arg	C
T	TAA ile	TGA thr	TTA asn	TCA ser	A
	TAG ile	TGG thr	TTG asn	TCG ser	G
	TAT ile	TGT thr	TTT lys	TCT arg	T
	TAC met	TGC thr	TTC lys	TCC arg	C
C	CAA val	CGA ala	CTA asp	CCA gly	A
	CAG val	CGG ala	CTG asp	CCG gly	G
	CAT val	CGT ala	CTT glu	CCT gly	T
	CAC val	CGC ala	CTC glu	CCC gly	C

term = termination code

Compare this code table with the more commonly produced table for mRNA codons and check that you understand the relationship between the two.

Fig. 17. The genetic code for DNA.

(d) Synonyms usually share the first two bases in their code and differ only in the last. The synonyms for proline, for example, all start GG.

(e) With very rare exceptions, it is universal. From bacteriophage to human being the genetic code is the same.

(f) It has full stops. These are the codes marked 'term' in the table. 'term' stands for 'terminator' and each of the terminator codes signifies that the end of a chain of amino acids has been reached and the process of protein synthesis can stop.

(g) It also has an instruction 'start reading here'. It may not be obvious which code word this is but it turns out to be either the code word for methionine or one for valine.

Fig. 17 shows the complete genetic code, expressed as the amino acids (or 'term') specified by all the possible code words in DNA.

How proteins are synthesised

The process by which the code in DNA is deciphered and translated into a sequence of amino acids can best be described in a series of diagrams. But before looking at them you will need to know about some of the participants in the scheme. A summary of the whole process from DNA to finished protein will help to introduce these things in context:

1. One strand only of the relevant part of the DNA is used as the coded message.

2. First the message is transcribed (copied) into a single stranded piece, not of DNA, but of its close relative RNA (ribonucleic acid). This transcribed piece of RNA is referred to as messenger RNA.

3. The messenger RNA attaches itself to ribosomes. These cell organelles act as sites for the assembly of the protein.

4. The individual amino acids which are going to make up the

protein are brought to the ribosomal assembly sites by another type of RNA, called transfer RNA.

5. At each ribosome site the message, which now resides in the messenger RNA, is translated into protein by the incorporation of the correct sequence of amino acids into a single protein molecule.

6. When the protein is finished it is free to leave the ribosome.

In that summary you will have come across the terms: messenger RNA, transfer RNA, ribosomes and the processes referred to as transcription and translation. What are these things?

RNA – ribonucleic acid

Ribonucleic acid is very similar to DNA but differs in the following important ways:

1. It contains the five carbon sugar ribose in place of deoxyribose.

2. It does not contain the base thymine. In place of thymine it has a similar base called uracil (though see tRNA).

3. It does not usually form a double stranded structure like that of DNA.

In other structural respects DNA and RNA are alike.

Base pairing and RNA

The specific base pairing rule still applies in the case of RNA, though it needs to be modified on account of the substitution of uracil for thymine in RNA. Uracil (U) pairs with adenine, just as thymine does.

RNA can, in rare instances, be double stranded through complementary base pairing. It can hybridise with single stranded DNA if the two strands are complementary. A single strand of RNA can bend back on itself to form a double stranded region of its molecule, using the base pairing mechanism.

The rules of specific base pairing are of paramount importance during transcription and translation of the genetic code.

Messenger RNA

In eukaryotic cells (cells of higher organisms that have proper nuclei and membrane bound cell organelles) the DNA, carrying the genetic code, is in the nucleus, whereas the sites of protein synthesis, the ribosomes, are in the cytoplasm. A double membrane, surrounding the nucleus, separates the two. The job of messenger RNA (mRNA for short) is to convey the code message from the DNA in the nucleus to the ribosomes. In bacterial cells (prokaryotic cells) there is no proper nucleus and no nuclear membrane. But mRNA does the same job of conveyance.

▶ *Key point* – Messenger RNA carries the genetic information from the DNA to the sites of protein synthesis.

Synthesis of messenger RNA

The synthesis of mRNA is equivalent to the synthesis of a new DNA strand in DNA replication. In both cases one strand of the DNA molecule, having been separated from its partner strand, acts as the template for the new strand, whether it be DNA or RNA. Therefore, just as the old and the new strand of newly replicated DNA are complementary, so mRNA is complementary to one of the strands of the DNA which made it (see fig. 18).

▶ *Key point* – Messenger RNA carries the same information as the gene which makes it, because it is complementary to the DNA comprising the gene.

The synthesis of mRNA by a gene is called transcription.

Transfer RNA

Once mRNA is bound to ribosomes the actual translation of the code (converting the message into protein) can begin. What this entails is bringing the correct amino acids in the correct sequence to the site of protein synthesis, where they can combine together. Carrying amino acids is the job of transfer RNA (tRNA for short).

Each amino acid has its own specific tRNA. All tRNA species are short (about 80 nucleotides long) pieces of RNA which have a characteristic three dimensional structure brought about by folding of the molecular strand and the stabilising of the folds by base pairing. An interesting feature of tRNAs is that they contain unusual bases, in

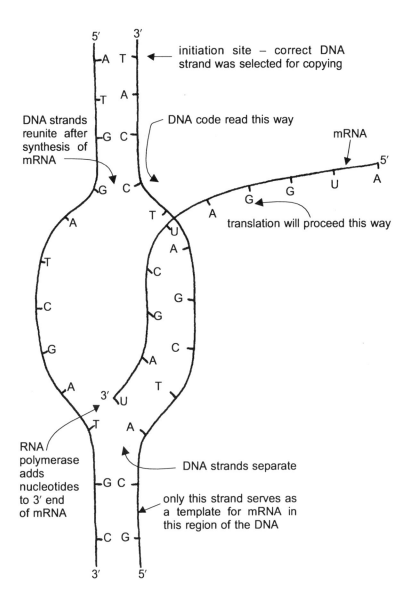

Fig. 18. Transcription: the synthesis of mRNA. RNA chain growth proceeds at a rate of approximately 40 nucleotides per second in E. coli under optimum conditions.

addition to the normal A, G, C and U of RNA. They can even contain the base thymine, an exception to the rule that thymine is found in DNA but not in RNA.

Key functional properties of tRNA
From the point of view of function the most important properties of tRNAs are that they:

(a) combine with amino acids

(b) do so specifically – each type of tRNA combines with only one particular amino acid – so each tRNA can be named after its amino acid partner (alanine tRNA, phenylalanine tRNA, etc.)

(c) are attracted to and bind to ribosomes

(d) include in their structure a triplet of bases which is complementary to the triplet of bases on mRNA that codes for the amino acid they carry

▶ *Key point* – Each tRNA carries its own amino acid to the site of protein synthesis and inserts it in the correct place in the protein molecule because it recognises the matching code word on the mRNA.

Codon and anticodon

These are two other terms that need explaining. The sequence in the synthesis of protein is as follows:

code in DNA $\longrightarrow$ code in mRNA $\longrightarrow$ recognised by triplet of bases in tRNA $\longrightarrow$ correct amino acid inserted

For example: AGA $\longrightarrow$ UCU $\longrightarrow$ AGA $\longrightarrow$ serine

You will notice that, at each step, the triplet of bases changes to its complementary form. The triplet on tRNA (AGA) can only recognise the mRNA code (UCU) for the very reason that it is complementary and can, therefore, hybridise (base pair) with it. This is the essential trick which brings the right amino acid into play and allows the code to be translated. The triplet in mRNA is called the codon and the triplet in tRNA is called the anticodon.

▶ *Key point* – mRNA codon and tRNA anticodon are complementary triplets of bases.

Messenger RNA is complementary to the DNA that forms it. The anticodon on tRNA is complementary to the codon. Therefore the triplet code word on DNA that specifies an amino acid is more like the anticodon. But note that it might not be the same because of the substitution of U for T in RNA.

Ribosomes
Most of the RNA in both eukaryotic and prokaryotic cells is found associated with protein in small organelles called ribosomes. Prokaryotic ribosomes, such as those found in bacteria, are about 20 nanometres (20 millionths of a millimetre) in diameter and contain roughly twice as much RNA as protein. Eukaryotic ribosomes are slightly larger and contain a larger proportion of protein. A single cell of E. coli may contain anything from a thousand to fifteen thousand ribosomes.

E. coli ribosomes have two parts, one smaller than the other. The smaller part is attracted to and binds with mRNA during protein synthesis. The larger unit can attach to two tRNAs simultaneously, one in the process of adding a new amino acid to the growing protein chain, the other bringing the next amino acid in the sequence. By attaching to both mRNA and tRNAs, the ribosomes bring all the players in the synthesis of protein together, in the same place, at the same time.

▶ *Key point* – Ribosomes bind to both mRNA and tRNA and are the sites of the assembly of amino acids into protein.

Diagrammatic account of protein synthesis in E. coli (fig. 19)
One thing to bear in mind when following the diagrams is that practically all the details of protein synthesis were established using the bacteria E. coli. That is not to say that the details do not apply in other organisms, but it is necessary to be careful when extrapolating findings from one system to another. For example in E. coli the process

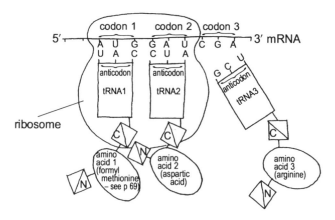

(i) The first two amino acids in the chain are brought together by the binding of their respective tRNAs to the ribosome. The anticodons on the tRNAs recognise and pair with the mRNA codons. The third amino acid, attached to its tRNA, waits in the wings.

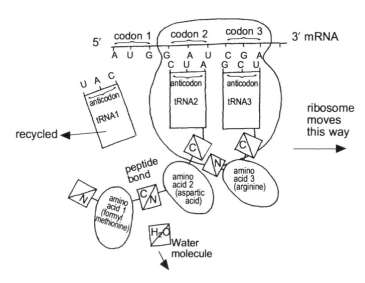

(ii) The carboxyl group of amino acid 1 and the amino group of amino acid 2 form a peptide bond and are, thereby, attached together. In the process a water molecule is lost. The rRNA that carried the first amino acid is recycled. The ribosome moves along the mRNA as far as the next codon and the tRNA carrying the third amino acid binds with it.

Fig. 19. Protein synthesis in E. coli – the translation of the genetic code.

of transcription can start while the process of translation into mRNA is still continuing. This is possible in a prokaryotic organism because nothing separates the DNA from the ribosomes, but it would not be likely to happen in a eukaryotic cell where the nuclear membrane separates the two processes.

Questions about the genetic code

1. · We know that only one of the strands in DNA carries the genetic message – but which one?
2. In which direction is the DNA code read?
3. How does the cell know where to start translating the code for one protein?
4. How does it know when it has come to the end?

Answers

1. In some cases both strands of the DNA act as the template for the synthesis of mRNA, but not, of course, in the same place. The rule still applies, that the genetic code resides in only one strand of the DNA in any one gene. It is not clear how the 'correct' strand is identified.

2. The code on mRNA is read from the 5′ end. This means that the code on DNA must be read in the opposite direction, that is from the 3′ end (see fig. 18).

3. A type of tRNA was discovered by Frederick Sanger and his colleagues that carried the amino acid N-formylmethionine. The odd thing about this amino acid is that it can only attach to another amino acid on one side, via its carboxyl group. It can, therefore, never be incorporated into a chain of amino acids making up a protein – unless it is at one end. This happens to be the end at which synthesis of proteins begins. So protein synthesis starts with the attachment of N-formylmethionine tRNA to mRNA on a ribosome. The codon for this species of tRNA is either AUG or GUG. These codons correspond to TAC and CAC on the DNA.

4. The so called nonsense codons or terminator codons, UAA, UAG, and UGA signify an end to protein synthesis. They were originally referred to as nonsense codons simply because they did not translate into any amino acid. They are the full stops in the genetic message.

How gene function is controlled

Every gene has the capacity to produce a protein by means of the process we have just looked at, and to keep on producing proteins by that means. But not every gene of an organism is doing so at any one time. It would be ridiculous if this were the case. Proteins would be made whether or not they were required and in amounts that bore no relationship to the needs of the cell. There must be a control system that switches genes on and off and regulates which proteins are made, when they are made and in what amounts they are made.

Reasons for a control system
This is vital for at least three reasons:

1. *Efficiency*
It is wasteful of resources for a cell to produce proteins that it does not need. Control of gene function leads to efficiency.

2. *Specialisation*
In multicellular organisms control of gene function allows for specialisation of cells. Only cells of the pancreas, for example, produce the protein insulin even though all other cells of the body carry the gene that codes for insulin. Each cell type produces different proteins because different genes are switched on in each case.

3. *Development*
Development depends on different systems coming into operation at different times. A seed, for example, needs enzymes to metabolise its food reserves during germination. Only later does it need the enzymes of photosynthesis, and later still it makes the proteins associated with reproduction. The right genes come into operation at the appropriate time leading to a proper developmental sequence.

▶ *Key point* – In any one cell at any one time only a small part of the genetic information is being translated into protein. Some genes are switched on. Most are switched off.

Lactose metabolism in E. coli
The first mechanism for regulation of gene function to be described in molecular terms was the lactose utilising system in E. coli. When E.

(a) lactose absent

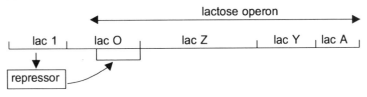

regulatory gene makes a repressor substance	repressor combines with operator and blocks the synthesis of mRNA (switch off)	all genes repressed – no mRNA synthesised no enzymes produced

(b) lactose present

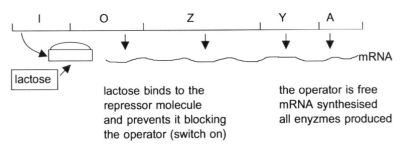

lactose binds to the repressor molecule and prevents it blocking the operator (switch on)	the operator is free mRNA synthesised all enyzmes produced

lac I =　regulatory gene

lac O =　operator – part of the DNA responsible for the initiation of mRNA synthesis for the whole operon – a sort of on/off switch

lac Z =　gene responsible for making the enzyme galactosidase

lac Y =　gene responsible for making a permease enzyme, without which lactose cannot enter the bacterial wall

lac A =　gene making a third enzyme associated with lactose metabolism

Note:

(i)　The three genes associated with lactose metabolism are found together on the bacterial DNA. They are transcribed as a unit, giving a single piece of mRNA.

(ii)　In the absence of lactose none of the genes makes mRNA coding for any of the three enzymes. The whole system is switched off.

(iii)　In the presence of lactose the system is switched on and all three genes synthesise mRNA.

Fig. 20. The lac operon regulating system in E. coli.

coli is given lactose as its only source of energy, it first breaks it down to the simple sugars, glucose and galactose, by means of the enzyme β-galactosidase. This enzyme is an example of an inducible or adaptive enzyme. It is only produced if its substrate, lactose, is present. In the absence of lactose the gene that makes β-galactosidase does not operate. It is said to be repressed.

The action of lactose as an inducer of an enzyme responsible for its metabolism avoids the unnecessary synthesis of the enzyme at times when it would have no work to do. In this way the bacterium conserves its resources.

The control system for the genes associated with lactose metabolism in E. coli was worked out by Jacques Monod and others by 1961. The system is summarised in the diagram (fig. 20). Note that regulation is at the first stage of protein synthesis, namely transcription of the genes into mRNA. By switching genes off at source the greatest saving is made.

Gene repression

Just as genes can be switched on or induced by the right substrate so they can be switched off or repressed by a gene product. As an example we'll look at the control of tryptophan synthesis (fig. 21).

It is far more efficient for a bacterium to utilise readily available materials than to make them for itself. So it would seem a good idea if the presence of an outside source of tryptophan repressed the formation of any of the enzymes involved in the synthesis of tryptophan. This is the principle of enzyme repression.

Gene control in eukaryotes

Operon systems

No operon systems quite like those found in bacteria have been discovered in higher organisms. This is not surprising considering the added complexity of genes being parts of chromosomes and confined to the nucleus of cells in eukaryotes.

Substrate control

This type of control, much like the β-galactosidase switch in E. coli, depends on enzymes being induced by their substrate, and is known to work in yeast cells. Here, for example, the enzymes associated with galactose fermentation are only produced when the cells are supplied with galactose. In the absence of galactose, no such enzymes are

(i) tryptophan absent

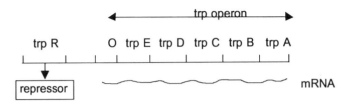

regulatory gene makes a repressor which cannot, by itself, block the operator and prevent mRNA being made

all enzymes are produced and tryptophan is synthesised

(ii) trytophan present

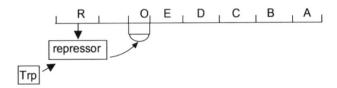

tryptophan and repressor, together, bind to the operator and prevent the synthesis of mRNA

no enzymes are made and tryptophan is not synthesised

Key
O   =   operator
trp R =   regulatory gene

trp E, trp D, trp C, trp B and trp A = five genes making enzymes associated with the reactions leading to the synthesis of tryptophan (see fig. 15).

Fig. 21. Gene repression – tryptophan synthesis in E. coli.

formed. Only the details differ from the lac operon model for β-galactosidase in E. coli. The principle is the same.

Hormone action

Hormones have the power to switch on whole banks of genes simultaneously. Steroid hormones enter cells and bind to specific receptor proteins. The hormone–protein complex is then capable of switching on the transcription of the genetic code found in specific genes. Developmental mechanisms can often best be explained on the basis of a hormone operated gene switch. The moulting hormone of insects, for example, is capable of turning a larva into an adult, a transition requiring the activity of a whole new set of genes.

Tutorial

Progress questions

1. In what general ways are the structures of DNA and protein molecules alike? How do these similarities of structure suggest a way in which information in DNA could be translated into protein?

2. Explain what is meant by the term 'genetic code'.

3. Using the code table, work out what sequence of amino acids is synthesised when the following piece of DNA is read:

    TAC GAA TGA CAC TTT ACC ATT

4. In molecular genetics how is a gene defined?

5. What are:
    (a) transcription
    (b) translation
    (c) mRNA
    (d) tRNA
    (e) ribosomes
    (f) codon
    (g) anticodon?

Seminar discussion

1. How is the activity of genes controlled?

2. Can you suggest why the only amino acid not found in the tryptophan synthetase A protein is tryptophan itself? Think about the circumstances in which a bacterium would need to make tryptophan. It may also help to refer to the next chapter.

Practical assignments

Find out how the code was cracked. In particular find reference to the so called triplet binding assay.

Study tips

1. Learn the list of amino acids with their three letter abbreviations. This will not only help you answer questions but will speed up your reading.

2. Also learn, if not all the code words for the amino acids, then at least some of them, including the start and stop codes and examples showing characteristics of the code, such as the existence of synonyms.

3. To remind yourself which of the DNA bases is missing from RNA and replaced by uracil, note that U immediately follows T in the alphabet. Or think of vowels and consonants pairing in RNA (A with U, C with G).

4

Bacterial Genetics

One minute summary – The bacterium Escherichia coli has been extensively used in genetic research. It has contributed much to our knowledge of gene function and control. Bacteria have short life cycles, and can be cultured in very large numbers. Compared with higher organisms they have relatively accessible genetic systems. These are the reasons they have proved valuable research organisms. Genetic exchange can occur in bacteria in three ways: pure DNA can be taken up by a bacterial cell, two bacterial cells can become physically joined, and one cell can donate some of its genetic material to the other; or genes can be transferred from one bacterium to another by a virus acting as a carrier. In this chapter we will explore:

▶ the nature of bacteria
▶ genetic characters of bacteria
▶ selective media and how they are used
▶ genetic exchange in bacteria
▶ the conjugation process
▶ making genetic maps using bacterial mating experiments
▶ gene transfer by plasmids and the problem of multiple drug resistance

The nature of bacteria

Although this chapter will concentrate on bacteria as organisms for genetic studies, it is necessary to include some background knowledge of the structure, metabolism, nutrition, mode of growth and characteristics of bacteria.

Escherichia coli

By far the largest amount of research in bacterial genetics, has used the common gut bacterium, Escherichia coli (E. coli). Of all genetic organisms, E. coli is probably the one to which we owe most of our knowledge of molecular genetics. Consequently the chapter will focus almost entirely on this species.

One thing to bear in mind concerning the generalisations inherent in genetics is that they often originate from the study of single organisms. For example, much of what we know of the process of protein synthesis and of the genetic code comes from work with E. coli. That is not to say that it is not valid for other organisms. But it is always a good idea to be cautious in extrapolating specific findings to all situations.

▶ *Key tip* – When answering questions and writing essays, always quote the relevant organism to which your information refers.

Structure
Bacteria are prokaryotes. Below are listed the main differences between prokaryotic cells and eukaryotic cells. All higher organisms – from protista, algae and fungi to flowering plants and vertebrates – are eukaryotes.

	Prokaryotic cells	*Eukaryotic cells*
Size	0.5 – 2.0 microns.	20 – 200 microns (typical sizes).
DNA	Uncombined with protein. Plasmid DNA may be present.	Combined with protein in chromosomes. Some DNA in cell organelles.
Nucleus	No membrane bounded nucleus.	Nucleus present, separated from the cytoplasm by a double membrane.
Cell wall	Present, but not cellulose or fungal cellulose.	Present in plants (cellulose) and fungi but not animals.
Cell division	Binary fission. No mitosis.	Binary fission preceded by mitosis. Also reduction division (meiosis).
Cell organelles	No membrane bound organelles. Smaller ribosomes	Membrane bound organelles, *eg* mitochondria, chloroplasts. Larger ribosomes.
Exchange of DNA	By transformation, conjugation and transduction.	By sexual fusion of cells and recombination at meiosis.

Bacterial DNA

The DNA of E. coli is a single piece with its ends joined to make a circle. It may sometimes, loosely, be referred to as a bacterial 'chromosome'. The DNA of a bacterium, however, is not combined with proteins as is the DNA of the chromosomes of eukaryotic cells.

There are some four million base pairs in the DNA of E. coli. This represents about four thousand separate genes. This is of the order of one thousandth, or less, of the amount of DNA in a single mammalian cell. A piece of DNA containing four million base pairs has a total length of between one and one and a half millimetres. This may not sound much, but remember, it all has to fit inside a cell which is no longer than two micrometres long (two thousandths of a millimetre).

Plasmids

Bacteria may also contain what are known as plasmids: small lengths of DNA which are independent of the main DNA complement but which replicate in the same way. Like the bacterial chromosome, plasmid DNA is circular in form.

The DNA of some plasmids can become incorporated into the bacterial chromosome so that it becomes an integral part of it. This integration is not necessarily permanent and may be reversed so that the plasmid DNA is free again. Plasmids, therefore, can lead a nomadic life, sometimes part of the main chromosome, sometimes independent.

An example of a plasmid is the F factor which is associated with a sexual process in bacteria.

Growth and nutrition

E. coli is saprophytic. This means it obtains food by absorbing it from the surrounding medium. Its synthetic capacity is excellent. Given a single source of carbon, such as a simple sugar, and all the necessary inorganic ions, it can make all the organic molecules it needs for growth. This is an important point to remember in relation to some of the genetic characters we will come across. There is no single specific organic compound that E. coli needs from the nutrient medium on which it grows.

In genetic terms this means that the normal E. coli cell has all the genes to make all the enzymes, to synthesise all the materials it needs for growth.

▶ *Key point* – Organisms which have a full complement of genes, all in working order are referred to as 'wild type'.

Growth media

E. coli is normally grown either in an aerated nutrient broth or on the surface of a nutrient agar jelly. Nutrient growth media have a number of organic molecules in them. The bare minimum requirements for growth, however, are water, one source of carbon, such as glucose, and mineral nutrients, such as nitrate and phosphate. Wild type E. coli are perfectly happy with these basic requirements. A medium containing only the minimum nutrient requirements is called a minimal medium.

Many other things can be incorporated into growth media, either to allow the growth of bacteria that need additional nutrients, or to select certain types of bacteria according to their ability to utilise different substrates or their tolerance of harmful additives. Some media are referred to as indicator media because they identify certain types of bacteria by their responses to substances in the medium.

Growth media which are used to distinguish different genetic types of bacteria are called selective media. You will need to become familiar, firstly with the principle behind the use of selective media and, then, with some of the more commonly used selective media and their use.

Reproduction

In ideal conditions the number of individuals in a growing population of E.coli can double in about twenty minutes. In this time every cell has replicated its DNA and divided into two new cells, exact copies of the first. From one starting cell, in forty minutes there will be four bacteria, in an hour, eight and so on. In less than seven hours the number of bacteria is up to over a million. When grown on solid medium, bacteria form colonies. A colony is a mass of genetically identical individuals derived by cell replication from a single antecedent.

Genetic characters of bacteria are usually scored on the basis of the presence or absence of colonies growing on agar, or, sometimes, on the appearance of these colonies. It is important to realise two things about this method of working:

(a) In bacterial genetics it is populations that are being assessed.

(b) This method only works because bacterial reproduction gives rise to clones of millions of identical cells.

▶ *Study tip* – Refer back to chapter 1, 'The simplest form of inheritance'.

Genetic characters of bacteria

In order to do any genetic experiments it is first necessary to have what are known as genetic markers – that is, characters which differ from one individual or population to another. The type of characters used in bacterial genetics are quite different from those that feature most frequently in classical genetics. Bacteria are far too small for individual morphological characters to be of any value.

The types of genetic markers most commonly used in bacterial genetics are physiological. For example:

(a) specific nutrient requirements for growth
(b) ability to metabolise certain substrates
(c) sensitivity or resistance to harmful substances such as antibiotics
(d) sensitivity or resistance to infection by bacterial viruses (phages)

Note that all these characters are easily translated to presence or absence of colonies on an agar medium. Look at the table and see how different selective media distinguish between different genetic types of bacteria simply on the basis of whether they grow or not.

Example of selective media	Types of bacteria that can form colonies	Types of bacteria that cannot form colonies
Minimal	Wild type	Any nutritionally deficient type
Minimal plus leucine	Wild type and leucine requiring types	Any nutritionally deficient type – except leucine requiring ones
Lactose as only carbon source	Able to metabolise lactose	Unable to metabolise lactose
Presence of the antibiotic streptomycin	Streptomycin resistant	Streptomycin sensitive
Presence of T-one phage	T-one resistant	T-one sensitive

Genotype and phenotype

One of the advantages of using bacteria for genetic research is that they have only one set of genetic material (except that, during active growth, the DNA is always in the process of being replicated). For each genetic character there is only one gene (or, if you want to be pedantic, either one gene or one gene and one identical copy of that gene). This means that genotype and phenotype are directly and very simply related. To all intents and purposes they are the same thing.

Nomenclature

The phenotype of a bacterium is given a three letter abbreviation to which is added a superscript to denote the status of the character:

Lac^+ = able to metabolise lactose
Lac^- = unable to metabolise lactose
Leu^+ = able to make leucine; therefore not requiring it
Leu^- = unable to make leucine; therefore leucine requiring
Str^s = sensitive to streptomycin
Str^r = resistant to streptomycin
Ton^s = sensitive to the bacteriophage T-one
Ton^r = resistant to the bacteriophage T-one

The genes that control the phenotype are given the same notation except that the first letter is lower case. So, for example a gene that allows the metabolism of lactose would be lac^+, one that permits the synthesis of leucine would be leu^+, and so on. You will need to become familiar with this system of shorthand. One easy thing to remember is that $^+$ signifies wild type and is a positive property (can use lactose, can make leucine, etc.) while – signifies the negative, mutant form (cannot).

Genetic exchange in bacteria

▶ *Key point* – The term genetic exchange simply means the introduction of DNA from one bacterium into another and the subsequent shuffling of the different genes to produce a new combination. The process of swapping genes between different pieces of DNA is called recombination and the resulting, genetically modified individual is a recombinant.

Recombination in bacteria can result from three processes:

1. transformation
2. transduction
3. conjugation

You learnt about transformation in Chapter 2. Transduction will be mentioned in the next chapter. Here we will deal only with the third method of genetic exchange in bacteria, conjugation.

The first genetic cross

The first indication of a sexual process in bacteria came from an experiment carried out in 1946 by Joshua Lederberg and Edward Tatum. Fig. 22 shows a schematic summary of the experiment. The important thing to notice is that when two strains of E. coli are mixed they manage to produce a completely new genetic strain, that is, a recombinant type. This is proof of genetic exchange between the two original strains of bacteria. The recombinant progeny of the cross have inherited some of their genes from one parent strain and some from the other.

Some points about the experiment

(a) You should already understand, from what has gone previously, the nomenclature for the genetic types of the parents and progeny of the cross. Bio refers to the vitamin, biotin, which is a growth requirement for the strain A parent.

(b) The cells were washed, after being mixed, to remove all the complete nutrient medium. If any of this had been carried onto the minimal agar, there could have been some growth of the parent types.

(c) The minimal agar, in this case, is a selective medium, since it is selecting only those recombinants that require no growth factors. Neither parent can grow on minimal agar.

(c) Both parents were deficient in two ways, that is, they each required two growth factors. This is a way of ruling out the possibility of wild type bacteria, capable of growing on minimal agar, arising by mutation. The chance of two genes mutating simultaneously is minute.

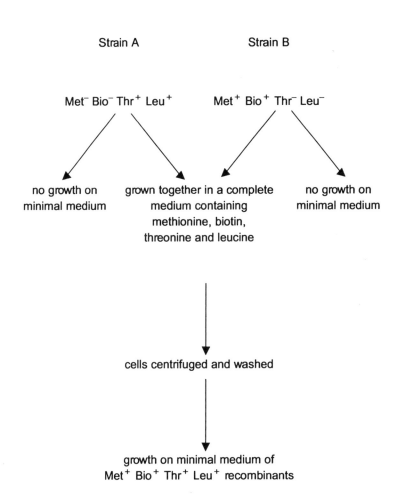

Fig. 22. Schematic summary of the first genetic cross involving bacteria.

(e) The result of the experiment could be explained away as an example of transformation. Subsequent experiments, however, showed that physical contact between parents was necessary for the production of recombinants.

The conjugation process

When two compatible bacteria come into contact with each other, they join by means of a conjugation bridge (also called a conjugation tube). The bacterial chromosome of one bacterium, the donor, moves across the conjugation bridge into the recipient . When conjugation is completed the recipient bacterium now has two copies of at least some of the bacterial genes and recombination can occur. The details of the process are shown diagrammatically (fig. 24). First, however, we need to know something of the types of E. coli that can engage in the conjugation process.

The F factor
A bacterium can only act as a donor of genetic material if it contains a plasmid called an F (for fertility) factor. The F factor can either be free in the bacterial cell, a small, independent circle of DNA, or it can be joined to the bacterial chromosome (fig. 23). In the latter case the bacterium is referred to as Hfr (high frequency), because recombinants are produced more frequently in crosses involving Hfr strains.

The merozygote
Conjugation ends with the recipient, F$^-$, bacterium in possession of two copies of all the genes that have been donated to it by the Hfr parent. It is very unlikely ever to receive a whole complement of genes, as the conjugation process rarely goes to completion. More usually the two joined cells break apart and only part of the donor DNA is passed across the conjugation bridge.

If the recipient, at the conclusion of conjugation, were to have two complete sets of genetic material it could be likened to the proper zygote that results from fertilisation in eukaryotes. Instead it is referred to as a merozygote or partial zygote. This is a similar situation to that which exists when a bacterial cell takes up pure DNA in the transformation process. The cell has all its own genes and, in addition, one or a number of 'floating' genes from another source.

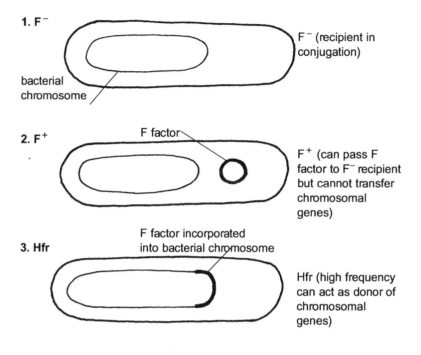

1. F⁻

bacterial chromosome

F⁻ (recipient in conjugation)

2. F⁺

F factor

F⁺ (can pass F factor to F⁻ recipient but cannot transfer chromosomal genes)

3. Hfr

F factor incorporated into bacterial chromosome

Hfr (high frequency can act as donor of chromosomal genes)

Compatible matings

(i) F⁺ x F⁻   The F factor is passed to the F⁻ recipient which becomes F⁺ as a consequence.

(ii) Hfr x F⁻   Some genes from the Hfr donor are transferred to the F⁻ recipient but the F⁻ cell does not become a potential donor itself.

Fig. 23. Bacterial mating types.

Hfr parent (donor)   F⁻ parent (recipient)
genotype = ABCDEFGH   genotype = abcdefgh

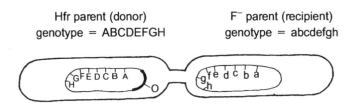

(1) The Hfr and F⁻ cells join by a conjugation tube.

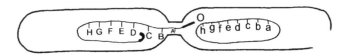

(ii) The Hfr chromosome breaks at the site of incorporation of the F factor. Part of the F factor (the origin: O) leads donor DNA through the conjugation tube into the recipient cell. The genes A-H follow in sequence.

(ii) At some stage the conjugation process is interrupted and the transferring DNA strand breaks. This may occur naturally or can be brought about by blender.

(iv) The cells separate. The recipient now has its own full complement of genes plus some genes from the donor. Note that the donated DNA and part of the recipient chromosome are homologous. Note also that genes G and H have failed to enter the recipient.

Fig. 24. The conjugation process in E. coli.

Recombination in the merozygote

The bacterial chromosome will only retain one copy of each of its genes. The question is – which one? Does it automatically keep the gene that has come from the donor bacterium? Or does it keep its own original one? We know it can't keep all its own genes. If it did, there would be no recombinants.

In fact, it may or may not keep the donated genes. They are kept only if they are recombined into the chromosome of the recipient. In other words, for a donated gene to be retained, the donor DNA has to be incorporated into the recipient DNA at the expense of the original gene or genes. The old DNA has to be swapped for the new. Any left over DNA, whichever parent it originates from, is discarded. Figure 25 shows what happens after conjugation.

Mapping the bacterial chromosome

It should be apparent from the diagrams showing the conjugation process that the bacterial chromosome, and the genes it carries, enter the recipient cell in a line, like the carriages of a train. Assuming the whole process takes a finite time (it actually can continue for many minutes), the temporal order of entry of different genes should correspond to the actual order of genes along the chromosome.

This is the principle used in mapping the genes on the bacterial chromosome through conjugation experiments. Mapping, as a general term, means establishing the linear order of genes and the relative distances between them.

The first genetic map for E. coli

Francois Jacob and Elie Wolman produced the first genetic map for E. coli in 1956. In fact, it was their success in doing so that led, ultimately, to the scheme for bacterial conjugation shown above.

To follow this experiment you will need to be sure of the nomenclature for bacterial genetic characters and of the role of different media in selecting particular recombinants. If you are not certain, read the relevant parts of this chapter again.

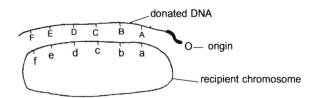

(i) The donated DNA pairs with the homologous region of the recipient chromosome.

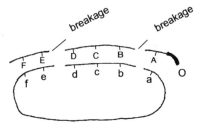

(ii) Both donor DNA and recipient DNA strands break in two places at homologous sites.

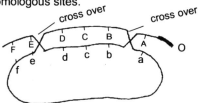

(iii) The DNA strands reunite, donor DNA to recipient DNA. This is known as crossing over.

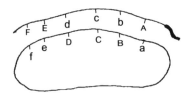

(iv) The intact chromosome has now exchanged the genes B, C, D though genes A, E and F are unaffected. The piece of DNA with the genes AbcdEF will be lost. Progeny of the recipient cell will all be of recombinant type: aBCDef

Fig. 25. Recombination in the merozygote. Refer to figure 24 to see how this situation arose.

An experiment of Jacob and Wollman on genetic transfer in E. coli

1. The following cross was made:

$$\text{Hfr Thr}^+ \text{ Gal}^+ \text{ Str}^s \text{ x F}^- \text{ Thr}^- \text{ Gal}^- \text{ Str}^r$$

2. The two strains were mixed and samples were taken from the mixture at various times.

3. The samples were agitated in a blender for a short while to separate the conjugating bacteria and were then plated onto two selective media.

4. The media used were:
    (a) minimal glucose agar containing streptomycin
    (b) minimal galactose agar containing streptomycin and threonine.

What recombinants were selected by the two media?

Medium (a) selected $\text{Thr}^+ \text{ Str}^r$ recombinants.
Medium (b) selected $\text{Gal}^+ \text{ Str}^r$ recombinants.

5. The number of recombinants growing on the two different media were recorded for the various times for which the mixture was incubated prior to blender treatment.

6. The results are shown in fig. 26.

Making the map

To obtain a recombinant of the type $\text{Thr}^+ \text{ Str}^r$ the thr^+ gene has to enter the recipient cell from the donor. This does not happen until eight minutes after mixing of the Hfr and the F^- strains. If conjugation is interrupted by the blender at any time before eight minutes, no recombinants of this type are produced.

The $\text{Gal}^+ \text{ Str}^r$ recombinants are not produced until conjugation has been continuing for twenty five minutes. In other words, whereas the thr^+ gene enters the recipient after eight minutes, the gal^+ gene has to wait another seventeen minutes before its turn comes.

We are now in a position to draw a map (see fig. 27).

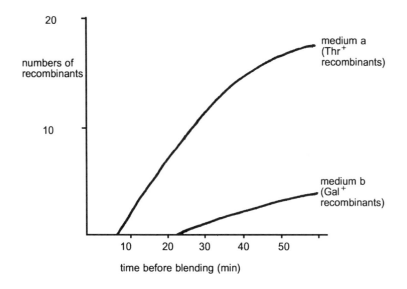

Fig. 26. Recombinants growing on selective
media after different incubation times.

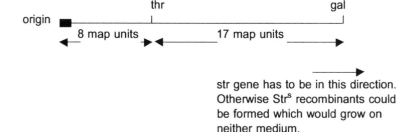

str gene has to be in this direction.
Otherwise Str^s recombinants could
be formed which would grow on
neither medium.

Fig. 27. Genetic map for thr and gal genes of E. coli. Note:
the map distance between thr and gal is likely to be the same
for different types of E. coli. The relative position of the two
genes with respect to the origin, however, will vary from one
Hfr strain to another. One map unit = 1 minute of
conjugation time.

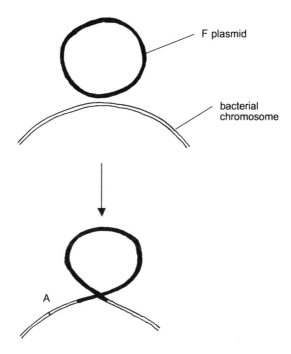

(i) The integration of an F plasmid into chromosomal DNA involves a single cross over between the two circular pieces of DNA.

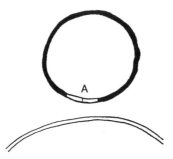

(ii) When next the plasmid becomes free from the chromosome it can take part of the chromosomal DNA with it (in this case the gene A).

Fig. 28. The transfer of chromosomal genes to a plasmid.

Gene transfer by plasmids

We have seen how the F plasmid can become integrated into the bacterial chromosome and thereby convert an F^+ cell to an Hfr type, capable of donating chromosomal genes to a recipient cell during conjugation. This integration is reversible. The F factor can spontaneously become free from the chromosome. When it does so it can take with it a portion of the chromosomal DNA. In other words the F plasmid can carry some chromosomal genes (see fig. 28).

During conjugation between an F^+ strain in which the F plasmid carries chromosomal genes, these genes are transferred along with the F factor. One consequence of this is that recipients in such a cross can, from then on, have two copies of some genes. One copy resides in their own chromosomal DNA, the other in the plasmid. The first instance of this phenomenon to be observed involved the lac gene in E. coli. A cross between F^+ Lac^+ and F^- Lac^- produced recombinants which were Lac^+ but, on close examination, were found to contain both lac^+ and lac^- genes. The phenotype of these partially diploid cells was Lac^+. In other words the lac^+ gene is dominant over the lac- gene (see Ch. 6).

Multiple drug resistance
A particularly important example of the transfer of genes attached to a plasmid is the case of the so called R or resistance factor. The R factor is a plasmid which carries, in addition to the fertility factor genes, a number of genes for resistance to different anti bacterial drugs. The fertility factor genes enable the plasmid to be passed from one bacterium to another, immediately rendering the recipients, and all their progeny, multiple drug resistant.

R plasmids are now commonplace in medical bacteriology and present a serious threat to the control of pathogenic bacteria. Indiscriminate use of antibiotics is a major cause of the spread of drug resistant bacterial strains.

Tutorial

Progress questions
1. In what form or forms is DNA found in bacteria?

2. Describe the phenotype of a bacterium which is:

$$\text{Lac}^- \text{Leu}^- \text{Str}^s$$

3. Why would the above bacterial type not grow on any of the following media?

    (a) minimal glucose medium
    (b) complete medium with lactose as the only carbon source
    (c) nutrient medium with streptomycin

4. What is meant by genetic recombination? In what ways can it be brought about in bacteria?

5. Describe the conjugation process, from the coming together of two genetically distinct bacterial cells to the production of recombinant types.

6. How can conjugation experiments be used to make genetic maps of bacterial DNA?

Seminar discussion
Is it a good thing to use antibiotics as a preventative measure in human medicine or routinely in animal husbandry? If not, why not?

Practical assignments
Find out what is entailed in the process known as replica plating and the value of the technique in genetic research using bacteria.

Study tips
1. Your understanding of bacterial genetics will greatly improve if you have some hands-on experience of culturing bacteria and using different media. If practical work of this sort is not a part of any course you are doing, try to get access to a microbiology lab or view a video on the subject.

2. Learn the most commonly referred to characteristics of bacteria used as genetic markers, with their three letter abbreviations. If you already know your amino acids this will help greatly.

5

Viruses and Viral Genetics

One minute summary – The bacterial viruses, known as bacteriophages, have been vitally important in contributing to our understanding of molecular genetics. Even more so than bacteria, viruses, with their rapid reproductive rate, can speed up the process of genetic research. Viruses have no metabolism of their own; they are entirely dependent on a host cell for their reproduction. It is the DNA or, in some cases, the RNA of a virus that is important in infecting a host cell, in controlling its activities and in implementing the reproduction of the virus itself. This very clear relationship in viruses between the genetic material and its function has made them particularly valuable as research tools. In this chapter we will explore:

▶ the general nature of viruses
▶ the structure and life cycle of bacterial viruses (bacteriophages)
▶ the value of bacteriophages in genetic research
▶ genetic characters of bacteriophages
▶ how genetic recombination can occur when two genetically different viruses infect the same bacterial cell
▶ how recombination happens
▶ lysogeny, prophage and transduction
▶ transduction for making detailed maps of bacterial DNA

What are viruses?

Viruses are halfway between living and non living. They have no independent metabolism, and can only reproduce by using the cellular mechanisms of other organisms. They are, in effect, obligate cell parasites. What makes them unique is that they are parasitic at the genetic level. They survive because their genetic material (DNA or RNA) can enter cells, suppress the role of the indigenous DNA, and implement its own instructions, using the resources of the host cell to do so.

Viruses are an order of size smaller than bacteria. The T-even bacterial viruses that have been widely used for genetic research, for example, are about two ten thousandths of a millimetre long. This is well below the limit of resolution of the light microscope.

Viral structure

The basic structure of all viruses is the same. A core of nucleic acid is enclosed within a protein shell. If you are reading this in conjunction with the account of the Hershey-Chase experiment (page 46) you will see that this experiment was only possible because of the simple structure of viruses and the clear division of the structure into two components, protein and nucleic acid.

There are both DNA and RNA viruses. RNA viruses are the only organisms which provide an exception to the rule that DNA is the primary carrier of genetic information in all organisms. There are also viruses in which the DNA is single stranded – another exception to a general rule. But viruses are exceptional things, and it should be borne in mind that none of them can reproduce without the genetic apparatus of other cells.

Bacteriophages

Just as most bacterial genetic research has used E. coli, practically all the genetic work done with viruses has been with bacteriophages. This chapter, therefore, from now on, will only consider this group of viruses.

The value of bacteriophages as genetic tools

Phages bring to genetic studies at least three important advantages:

1. They have a very short life cycle (of the order of thirty minutes).

2. They can be produced in vast numbers (concentrations of ten billion per millilitre are not unusual).

3. Their DNA is readily accessible.

To a lesser extent bacteria share these advantages, which accounts for the value of E. coli as a genetic organism. Between them, bacteria

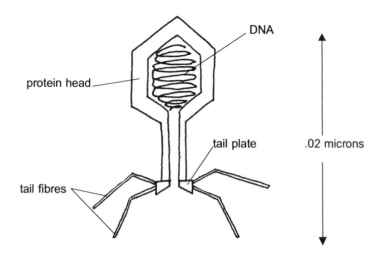

Fig. 29. The structure of T-even phage.

and bacteriophages, used as experimental organisms, have provided almost all of our present knowledge of molecular genetics.

Structure of T-even phages

The so called T-even phages (T for type, and 'even' because they are even numbered, T2, T4 and T6) belong to a group of viruses known as 'complex' viruses. As viruses go, their structure is quite elaborate (fig. 29), though it still represents the essential division into protein coat and nucleic acid centre.

Phage DNA

The T2 phage head contains about fifty thousandths of a millimetre of DNA, enough to code for one hundred and fifty average sized genes. This is not much more than five percent of the amount of DNA in the chromosome of E. coli, the phage's host.

One odd thing about the DNA of T-even phage is that it contains no cytosine. In its place there is the base hydroxy-methyl-cytosine – yet another exception to a general rule. But the base pairing rule still applies. The base HMC pairs with guanine, just as cytosine does.

The DNA of phage particles is released when the phage coat is disrupted. This can be achieved by quite simple treatments such as osmotic shock (suddenly transferring the phages from concentrated salt solution to water). The DNA is, therefore, easily extracted for analysis, or for experimental purposes.

Life cycle
When phage particles are mixed with susceptible bacteria, infection proceeds as shown in fig. 30.

One step growth curve
Infection of a suitable host bacterium by phage particles is followed by a period of about twenty five minutes, during which time, apparently, nothing happens. We now know, of course, and you can see in fig. 30, that, during this time, assembly of new phages is going on inside the host. The period is referred to as the latent period. If the number of infective phages in a mixture of E. coli and phage particles is counted, there is no change in the number during the latent period, but, as soon as the latent period is over, the number rapidly increases to nearly two hundred times the original count (fig. 31). Two hundred represents the burst size, that is the number of phages released on lysis of a single infected bacterial cell.

Genetic characters of phages

The huge numbers of phages that can be grown in a very short time vastly increases the possibilities of detecting rare mutant types. It is not difficult, therefore, to collect numbers of different genetic strains of phage with which to conduct genetic investigations. But what sorts of characters can phages show? They have no metabolism and they are far too small to observe, even with a microscope.

One class of genetic marker relates to plaque morphology.

Plaques
Plaques (pronounced 'plarks' or 'plaks') are the phage equivalent of bacterial colonies. They are clear, usually circular spaces on a culture of bacteria grown on a solid medium. The bacteria are grown so thickly that they form an uninterrupted covering on the agar surface, known as a bacterial lawn. Plaques are like holes in the lawn, formed by lysis of the bacteria within the plaque.

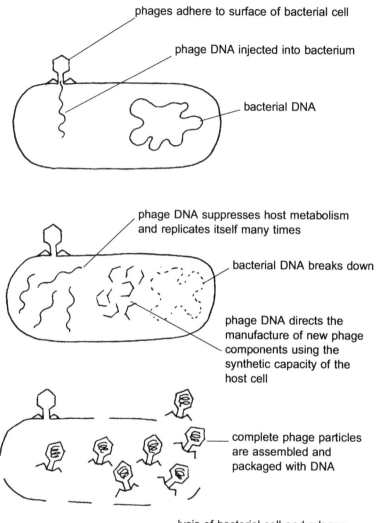

phages adhere to surface of bacterial cell

phage DNA injected into bacterium

bacterial DNA

phage DNA suppresses host metabolism and replicates itself many times

bacterial DNA breaks down

phage DNA directs the manufacture of new phage components using the synthetic capacity of the host cell

complete phage particles are assembled and packaged with DNA

lysis of bacterial cell and release of new phages

Fig. 30. The bacteriophage life cycle.

Just as a bacterial colony has originated from a single bacterium and, therefore, contains only one genetic type of bacterium, so a plaque originates from one infective centre (usually a single bacterium infected by a single phage) and can be reasoned to contain only one genetic type of phage. Reproduction of the phage from the first infected bacterium causes further infections, more reproduction, yet more infection and so on, spreading out in all directions until a visible plaque is formed.

Obtaining plaques
The procedure for obtaining plaques is as follows:

1. A dense liquid culture of E. coli (perhaps a billion bacteria per millilitre) is mixed with a suspension of phages so that there is approximately one phage to each bacterial cell.

2. The mixture is incubated for two minutes to allow for infection.

3. It is then diluted to one ten thousandth or less of its original concentration.

4. A drop of the diluted mixture is added to a small amount of liquefied agar medium containing uninfected E. coli. This is then poured evenly over a base of solid agar.

5. The uninfected bacteria grow to produce a lawn but, wherever an infected bacterium from the original mixture is encountered, all the bacteria in the vicinity are destroyed and a plaque is formed.

Plaque type mutants
One of the first phage mutant types to be recognised was the so called rapid lysis mutant of T2 phage. The normal (wild type) plaque formed by T2 on one particular strain of E. coli is small and round with a clear centre and an outer, misty halo. The rapid lysis mutant forms a larger plaque with a sharp edge and no halo. Rapid lysis mutants occur at a rate of about one in every thousand to ten thousand plaques.

Other examples of genetically mutant plaque morphology are: minute (very small plaques), turbid (misty plaques) and star (irregular shaped plaques).

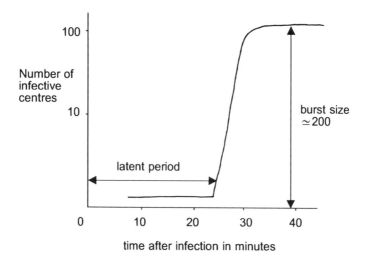

Fig. 31. One step growth curve for phage T4.

Host range mutants

Another type of mutant phage isolated early on in phage genetics was 'host range'. The name refers to the fact that host range mutants can infect a wider range of bacterial strains than the normal phage. In the same way that human beings can become immune to one type of influenza but are still susceptible to other types, bacteria can resist some phages but not others. Mutation of the phage can enable it, once again, to infect a previously resistant bacterium.

Recombination in phages

Once suitable genetic markers had been isolated in phages it became possible to test whether two genetically different phages could exchange genes; in other words, whether recombination could occur.

The evidence that, indeed, genetic recombination does occur in phages came from mixed infection experiments carried out in the late 1940s. Consider, for example, the following experiment using T2 phage and E. coli.

A culture of E. coli was infected simultaneously with two types of T2 phage: (1) host range and (2) rapid lysis. In terms of their genotype the two 'parents' in the cross were $h\ r^+$ and $r\ h^+$ respectively, where h is the host range gene and h^+ its wild type equivalent and r is the rapid lysis gene, r^+ its wild type equivalent.

Since the numbers of phages in the infection mixture were much higher than the number of bacteria, at least some bacteria received DNA from both types of phage. This amounted to the setting up of a potential cross:

$$h\ r^+ \text{ (host range) x } h^+\ r \text{ (rapid lysis)}$$

In the cross shown above, the production of any progeny of type $h^+ r^+$ or $h\ r$ (non parental types) would show that recombination occurs in a mixed infection. In fact a proportion of the progeny of such a cross were found to be recombinant. Genetic exchange had occurred.

Example of results of a cross between host range and rapid lysis mutants of T2 phage:

Progeny genotype	% all plaques	Type
$h^+\ r$	34	parental
$h\ r^+$	42	parental
$h^+\ r^+$	12	recombinant
$h\ r$	12	recombinant

A closer look at recombination

We have seen that recombination – the swapping of genes between DNA from different sources – can occur when two pieces of homologous DNA find themselves in proximity in a living cell. We saw it in the case of bacterial transformation. In the last chapter we saw it in the case of bacterial conjugation, and we have just seen it with bacteriophages. But what exactly happens when genes are exchanged? First, let's look at the conditions that need to apply before recombination can occur.

Conditions for recombination
1. *A living system*
Recombination is a controlled, living process. It can only happen in the presence of the necessary enzymes, a source of energy, and other resources, such as are found in a living cell.

2. *Homologous DNA*

Two pieces of homologous DNA are required. We must be absolutely clear, here, what is meant by homologous. It is not the same as complementary (referring to the two single strands of DNA that join together by base pairing). Usually, homology involves two pieces of double stranded DNA. Nor does it mean identical. Two pieces of DNA are homologous when they contain the same genes, in the same sequence, even if genes are mutant in one piece but not the other, or if genes are mutant in a different way or at a different site within the gene. Homology implies a largely similar, but not completely identical, sequence of bases or base pairs in two lengths of DNA. Only when homology in DNA is recognised can recombination occur.

3. *Proximity*

In order for two lengths of DNA to recombine, they must be in close proximity. This juxtaposing of molecules depends upon the recognition of homology, even to the extent of the direction of homology (the sequence of genes). A key property of DNA for the working of genetic systems is its ability to recognise homology, and for homologous regions of DNA to associate in parallel.

▶ *Example* – DNA introduced into a bacterial cell by transformation, if it contains bacterial genes, will associate with that part of the recipient chromosome where those genes are found. Similarly, the DNA entering a bacterium during conjugation will 'find' the homologous part of the chromosome of the recipient cell. If this were not the case, recombination could not occur in the controlled way that it does.

The mechanism of recombination

One reason that the general topic of recombination is being considered in this chapter is that the molecular mechanism of recombination was first looked for in phage genetics. Much of the evidence that has contributed to answering the question as to how recombination occurs has come from phage studies.

In the T2 phage cross we looked at, the situation inside the bacterial cells can be summed up as shown in fig. 32.

One possible way in which the h and r genes could be exchanged is for both pieces of DNA to break and for them to reunite in a different way. (See fig. 33.)

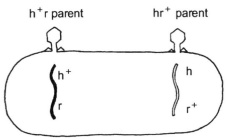

(i) Mixed infection by two different genotypes of phage results in the presence of two different but homologous pieces of phage DNA in a single bacterial cell.

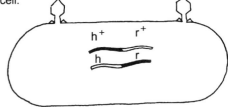

(ii) Recombinant DNA – the genes now united in a different way.

Fig. 32. Recombination in phage.

The scheme outlined in fig. 33 is now accepted as the mechanism for all recombination events in genetics, involving genes on homologous regions of DNA. It is known as 'breakage and reunion'. In phage it proved possible, by labelling one of the two pieces of parental DNA with heavy nitrogen (see Meselson and Stahl's experiment in chapter 2), to show conclusively that genetic exchange involves the physical exchange of DNA between the two parent DNA molecules. This, and other evidence established beyond doubt that when two pieces of DNA recombine they first break and then reunite in a new combination, as shown in the diagram.

It should be born in mind that the 'breakage and reunion' method of recombination is not, as the name might suggest, a random, accidental and uncontrolled process. A better description would be 'cutting, repositioning and repair'. Several enzymes are involved in the process and it is actually a quite complex piece of cell biochemistry.

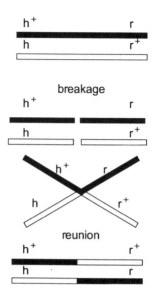

Fig. 33. Recombination by breakage and reunion:

Lysogeny and prophage

Some bacteriophages, known as temperate phages, when invading a bacterium with their DNA, do not necessarily bring about the lysis of the bacterial cell. Instead, the DNA of the virus becomes integrated into the bacterial chromosome in a manner similar to the integration of the F factor in Hfr bacteria. Once the viral DNA is part of the host chromosome, it replicates along with the bacterial DNA. In this way, as host cells reproduce, more and more bacteria carry the virus in its quiescent form. A bacterium carrying viral DNA in its chromosome is referred to as lysogenic and the viral DNA itself is called a prophage.

Reversal of lysogeny

A prophage is harmless to its host until such a time as it returns to a vegetative (infective) state. What happens then is that the phage reverts to the lytic cycle, reproducing itself and lysing the cell in which it has been residing. This spontaneous change, from prophage to virulent state, can be induced by stimuli such as ultra violet light.

Transduction

When a prophage reverts to the virulent state, a large number of new phages are assembled inside the host bacterium, before being released by lysis of the cell. These new phages may contain, as well as phage DNA, some bacterial genes derived from the host chromosome.

When the new generation of phages, in turn, infect other bacteria, the bacterial genes they carry are introduced into the new host cells. If one of these recipient cells is not, itself, immediately lysed by the infecting phage – either because it is lysogenic or because the phage is deficient in some way – the introduced genes can become part of its chromosome through recombination.

▶ *Key point* – In other words, genes carried by a bacteriophage vector can convert the genetic type of a recipient bacterium.
This form of gene transfer in bacteria is known as transduction.

Transductional mapping

In the last chapter we saw how a genetic map of a bacterial chromosome can be made from the results of conjugation experiments. You also learnt what a genetic map is. Transduction experiments can also be used to map the genes on a bacterial chromosome. Here we can only look at the basic principle involved.

Only a very small fraction of a bacterial chromosome can be transduced (carried from one bacterium to another by a phage vector). For a start there is only room inside a phage for a limited amount of DNA. In the phage of E.coli, called P1 phage, for example, the maximum amount of bacterial DNA carried by one transducing phage particle is about 2.5% of the bacterial chromosome. It follows, therefore, that only genes that are very close together on the chromosome will ever be co-transduced (simultaneously transferred from bacterium to bacterium by the same phage). If such genes are identified they can be placed close together on the map. The more frequently they are co-transduced the closer together they must be.

Use of transductional mapping

Transductional mapping has been used to obtain very detailed maps of small regions of bacterial chromosomes, most famously, perhaps, by Charles Yanofsky in his proof of the colinearity of gene and protein. This work, done in 1966, established the linear order (the map) of different mutant points within a single gene of E. coli by transduction experiments, and showed that this sequence mirrored exactly the linear order of amino acid changes in the proteins made by these mutants. In other words the gene and the protein it makes are co-linear.

You should note two points from Yanofsky's work:

1. The first is that it reinforces the relationship between genes and proteins established in chapter 3.

2. The second is that genetic maps need not confine themselves to the order and relative distance apart of genes but can delve further than that and map sites within a gene. Ultimately a genetic map can become a DNA map.

Tutorial

Progress questions

1. All viruses consist of two basic components. What are they?

2. The bacteriophage equivalent of a bacterial colony is a plaque. What is a plaque and how is it formed?

3. How does genetic recombination occur in bacteriophages?

4. Explain the meaning of the terms:

    (a) lysogeny
    (b) temperate phage
    (c) prophage
    (d) transduction
    (e) transductional mapping

Seminar discussion

1. Why have viruses sometimes been considered not to be living things?

2. Why are viruses particularly valuable in genetic research?

Practical assignment

A number of the exceptions to the general rules about the structure and characteristics of DNA and RNA are found in viruses. See how many such exceptions you can come across in your reading.

Study tips

1. Refer back to chapter 2 and make sure you understand the significance of the Hershey-Chase experiment in relation to the life cycle of a T even phage.

2. Look for good electron micrographs of viruses and, in particular, bacteriophages. These will really bring the study of viral genetics to life, especially pictures showing DNA extruded from the body of a bacteriophage or being injected into a bacterial cell.

6

Genetics of Higher Organisms I
Mitosis, meiosis and Mendelian inheritance

One minute summary – The modern science of genetics started with the work of Gregor Mendel on inheritance in the garden pea. Mendelian genetics forms the basis of the branch of genetics which involves the study of inheritance in higher organisms and is known as classical genetics. Although higher organisms were, historically, the first to be investigated, their genetics is made more complicated than that of bacteria and viruses by the organisation of the genetic material into chromosomes and by their diploid state. Mendel's principles are fundamental to an understanding of plant and animal genetics and are best learnt in relation to the behaviour of chromosomes at cell division. In this chapter you will learn:

▶ about chromosomes
▶ how chromosomes behave at mitosis and meiosis
▶ the genetic consequences of meiosis
▶ Mendel's monohybrid and dihybrid crosses and how the results are interpreted
▶ the meaning of the terms gene and allele in classical genetics
▶ what can cause exceptions to Mendelian ratios
▶ how some genes reside in the cytoplasm of cells

Classical genetics

Classical genetics is based on the study of inheritance in eukaryotic organisms. Historically, this should have been the first topic in a book on genetics. The scientific study of genetics began with higher organisms and, only later, progressed to work with bacteria and viruses.

There is a lot to be said, however, for taking advantage of the accumulated knowledge of all genetic research and – disregarding history – finding a logical progression from simple and basic to more complex. This is what this book tries to do. Molecular genetics and microbial genetics have come first for two reasons. Firstly, DNA is the basis of all genetics and, secondly, the genetics of microorganisms is a lot simpler than that of higher organisms. After all, it is to be expected that more highly evolved living organisms will have more complex genetic systems.

Differences between genetics of microorganisms and higher organisms

Two main differences between microorganisms and higher plants and animals make the genetics of the latter more complicated and less accessible to research:

1. The DNA of higher organisms is intimately associated with proteins in structures called chromosomes. Its role in heredity is, therefore, more difficult to resolve. You will remember, from what has gone before, that the DNA in bacteria and viruses is in the form of a single DNA molecule.

2. The life cycle of higher organisms invariably involves a stage during which there are two complete sets of genetic material in each cell. In most cases this is the most prolonged and dominant stage of the life cycle. The two sets are homologous but, usually, not genetically identical. This complicates the genetics of higher organisms considerably. Bacteria and viruses, on the other hand, normally only have a single full complement of genes.

Chromosomes

The DNA of eukaryotic organisms is found in chromosomes. Chromosomes, in turn, are confined to the nuclei of cells. A chromosome is approximately half by weight DNA and half protein. DNA and protein are combined in chromosomes in such a way that the DNA is not readily accessible. For example, it cannot be attacked by DNA degrading enzymes, except in certain regions. The disposition of DNA in chromosomes is, therefore, still not entirely clear. From the genetic point of view, however, it is fair to regard chromosomes as representing single, continuous DNA molecules, compressed from end to end by tight coiling.

The coiling of DNA in chromosomes is such that the linear arrangement of genes is not changed. In other words the order of genes along the chromosome is the same as the order of genes on the DNA molecule within the chromosome.

Active and non active states of chromosomes

During the non dividing stage of the cell cycle known as interphase, chromosomes are in their most elongated state. Only during this phase of the life of a cell can proteins be synthesised, DNA be replicated and genetic exchange between chromosomes take place. Actively metabolic chromosomes are not visible as discrete units in the cell. They are too extended. When cells divide, however, both by mitosis and at meiosis, chromosomes shorten dramatically and can be seen under the light microscope. In this state their metabolic functions cease.

Chromosome sets

The number of chromosomes per cell is a permanent and characteristic feature of an organism. For example, human cells have 46 chromosomes, those of fruit flies 8, cells of pea plants 14, and of the domestic dog 78.

Why are all these figures even numbers? It's because, in cells of most higher organisms, during the greater part of the life cycle, there are two sets of chromosomes. Each chromosome has a homologous partner (see page 102). Cells with two sets of chromosomes and, therefore, two sets of genetic information are said to be diploid. When a cell has only one set of chromosomes it is haploid. The haploid number is the minimum viable number of chromosomes.

Chromosomes at cell division

Genes lie within chromosomes. A knowledge of the behaviour of chromosomes at cell division and, more particularly, at meiotic reduction division, is therefore the best basis for understanding the inheritance of genetic characters in eukaryotes. Refer to chapter 1. There we saw how mitosis is the basis of cell division that results in perfect replication and inheritance of the genetic material (the simplest form of inheritance). Meiosis, on the other hand, halves the genetic content of cells and is the prelude, at some later stage in the sexual cycle, to cell fusion and the remixing of genes.

The diagrams (fig. 34) show the fate of a cell with two pairs of chromosomes, at mitosis and meiosis respectively. The chromosomes labelled A1 and A2 are homologous, as are the pair labelled B1 and B2. Make sure you understand the meaning of the term 'homologous' before following the diagrams. Note that A and B chromosomes can be distinguished by their different sizes and the positions of their centromeres. Centromeres are points on chromosomes which have an important role to play in the movement of chromosomes during cell division. The position of the centromere can vary from one chromosome to another but is always the same for chromosomes which are homologous.

Only the genes they carry can distinguish the homologous chromosomes A1 from A2 and B1 from B2.

Mitosis

1. *Replication*
Each chromosome is copied exactly. At first, the two chromosomes, the original one and the newly made one, are held together at the centromere. It can be shown that the replication of the DNA in chromosomes is a semi conservative process, as is the replication of DNA in bacteria (see page 38). Chromosomes that have just been copied and are held together in pairs are called chromatids.

2. *Assembly (metaphase)*
All the replicated chromosomes (pairs of chromatids) assemble in the centre of the cell. This stage of mitosis is called metaphase.

3. *Separation*
The centromeres that hold sister chromatids together divide, and each chromatid of a pair separates to a different side of the cell. In this way the two sides of the cell end up having a complete group of chromatids, one from every pair. It is important to realise that, during mitosis, the two chromatids that make up a pair are identical. It makes no difference, therefore, which one of a pair goes to which side of the cell.

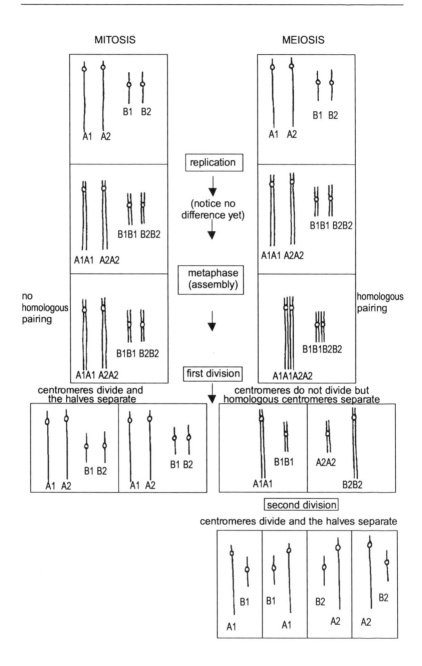

Fig. 34. Mitosis and meiosis.

4. Reconstitution of the nucleus

During mitosis the nucleus of the cell breaks down and the chromosomes are liberated into the whole space available. Following separation of the chromatids, if no mistakes are made, there should be the same number of chromosomes on each side of the cell. The chromosomes on one side are identical to those on the other. They are exact copies of the chromosomes from the original cell. The nuclei (there are now two groups of chromosomes, one for each new cell) reform and the net result of mitosis is two genetically identical cells.

Meiosis

You saw that in mitosis there is a doubling of the genetic material and a doubling of the number of cells. The amount of genetic material per cell, therefore, remains the same. In meiosis, on the other hand, there is also a doubling of the genetic material but a quadrupling of the number of cells. The amount of genetic material per cell is, therefore, halved. The number of chromosomes per cell is halved. This is why meiosis is referred to as reduction division.

Stages of meiosis

1. Replication

This is the same as in mitosis. The net result of meiosis is a perfect doubling of the genetic material, just as it is with mitosis.

2. Assembly (metaphase)

A feature of meiosis that distinguishes it from mitosis is that homologous chromosomes pair together. Pairing of homologous chromosomes is equivalent to the way in which homologous regions of DNA recognise each other and associate together. This was discussed in chapter 5. From the very beginning of meiosis, homologous chromosomes are paired in a highly specific way, lying closely together lengthways, homologous region to homologous region.

Don't forget that by the time chromosomes are visible, during both meiosis and mitosis, they are replicated into matching chromatids. So paired homologous chromosomes in meiosis are actually paired pairs of chromatids. At metaphase of meiosis, with all the chromosomes assembled across the middle of the cell, what is seen is associations of four chromatids.

3. Separation

Here we come to a big difference between mitosis and meiosis. In mitosis the centromeres divide, sister chromatids separate, and that, more or less, finishes the process. In meiosis the centromeres do not divide – yet. Instead they repel each other (remember there are two centromeres in an association of four chromatids because of homologous pairing) and carry two chromatids each to either side of the cell.

The result of this first division in meiosis is, as in mitosis, two cells with the same amount of genetic material as the original cell. But look at the diagram – the two cells, after one division, are not identical. One has chromatids labelled A1, A1, B1 and B1, the other has A2, A2, B2 and B2. Already we see a source of genetic variation.

4. Second division

Meiosis does not stop at the end of the first division, where mitosis stops. Following the first division the centromeres – which are still holding sister chromatids together – divide, and the chromatids themselves separate. Two cells, the products of the first division, simultaneously, divide in this way. The final result of meiosis, therefore, is four cells with half the original chromosome number in each.

Notice, again, from the diagram, that the four products of meiosis are not all genetically identical.

Meiosis and heredity

Meiosis either gives rise to gametes directly, the four cell products of the meiotic division, themselves, developing into gametes, or it does so indirectly. In either case the events of meiosis determine exactly what genes from each side of a cross join together at fertilisation to produce the next generation. The events of meiosis are consequently of the utmost importance in determining inheritance.

▶ *Key tip* – If there is a single key to the understanding of classical genetics, it is in being completely familiar with the behaviour of chromosomes and genes during meiosis. You would be well advised to make sure you know and understand those features of the meiotic process that are of genetic significance.

The experiments of Gregor Mendel

Gregor Mendel is a seminal figure in the history of genetics. He established that the secret to interpreting heredity lay in:

1. looking at single character differences between individuals, rather than some vague generality of similarity and difference

2. treating his results mathematically, and understanding both the predictable and the unpredictable nature of chance

3. interpreting his results according to a particulate theory of inheritance in which discrete 'factors' are responsible for single phenotypic characters.

Mendel and meiosis
Mendel knew nothing of meiosis but, had he done so, he would have seen that his laws could most easily be interpreted in the light of the movement of chromosomes at meiotic cell division.

One small point to bear in mind in relation to the place of meiosis in explaining Mendel's work is that, in flowering plants, meiosis does not give rise directly to male and female gametes. You should refer, at this point, to an account of the life cycle of a flowering plant to see exactly what does happen. It is possible, however, to ignore the details and to imagine that gametes in flowering plants are the direct products of meiosis. This is what we shall do.

The experiments
Mendel used garden peas as his experimental material. He had access to pure breeding varieties (pure breeding means that they reproduce each generation without varying) that differed in distinct ways. For example, some were tall and some dwarf; some had round seeds and some wrinkled seeds; some had green seeds and some yellow ones.

The other important feature of peas that made them suitable for Mendel's purposes is that they normally fertilise themselves. Crossing two plants requires the deliberate manual transfer of pollen from one flower to another. Mendel was, therefore, in complete control of his breeding experiments.

Monohybrid crosses

Mendel's first crosses were ones in which the two parents differed by a single character. A cross of this sort is called a monohybrid cross.

His first experiment, for example, was a cross between a round seeded variety and a wrinkled seeded variety. He made the cross using the different varieties as both male parent (pollen donor) and, in separate instances, female parent (seed producer). Crosses between the same genetic types, that differ only according to which is the male parent and which the female, are called reciprocal crosses.

Whatever way he made the cross Mendel found that the result was always the same. The progeny were round seeds. The wrinkled seeded character seemed to have disappeared.

Dominant and recessive characters

Because all the seeds produced in this first cross were round, Mendel described the round character of seeds as the dominant character. He called the wrinkled character the recessive character. In every one of his monohybrid crosses, involving, in all, seven contrasting pairs of characters, Mendel found one character, in each case, to be dominant and to appear in the first generation of a cross to the exclusion of the other character. These findings are summarised in the table below.

Contrasting characters used by Mendel in his crosses

Dominant	Recessive
round seeds	wrinkled seeds
yellow seeds	green seeds
violet flowers	white flowers
inflated seed pods	constricted seed pods
green seed pods	yellow seed pods
flowers along the stem	terminal flowers
tall plants	dwarf plants

In each case the character on the left is the one that appears to the exclusion of the other in a monohybrid cross between pure breeding parents differing in the way described.

One possible area of confusion in relation to Mendel's characters is whether a seed character such as round or wrinkled is a character of

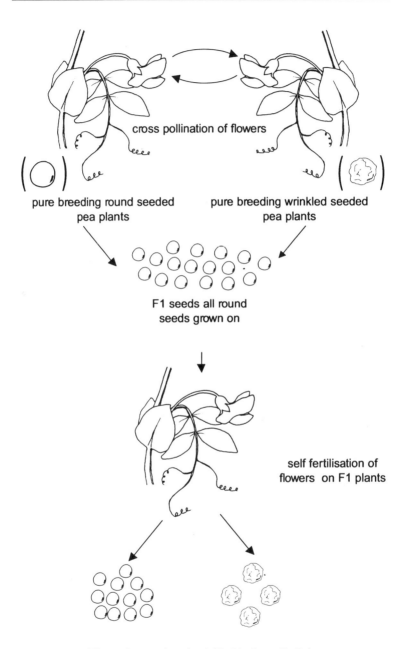

Fig. 35. Summary of Mendel's first monohybrid cross.

the plant that carries the seeds or a character of the next generation. Both the contrasting pairs of seed characters in the list above are, in fact, characteristics of the next generation. So, Mendel could record the result of a cross as soon as the seeds formed, without waiting for them to grow into plants.

From F1 to F2 generation

Mendel referred to his first generation of plants (or seeds) from a monohybrid cross as the F1 (first filial generation). He allowed the round F1 seeds to grow and the resulting plants to pollinate themselves in the normal way. The seeds formed by these plants were of both types, some round and some wrinkled. The wrinkled character, hidden in the F1 plants, had appeared a generation later in the F2 (second filial generation).

Mendel's genius was his recognition of a simple relationship between the number of round seeds and the number of wrinkled seeds in the F2. From a total of 7,324 seeds he had 5,474 round ones and 1,850 wrinkled, a ratio of 2.96:1. This, and other similar ratios Mendel obtained from the F2 of monohybrid crosses, was as near to 3:1 as made no difference.

Interpretation of the monohybrid cross

Production of the F1 generation

The monohybrid cross between round and wrinkled seeded plants to the F2 stage can be explained as follows:

1. The character of seeds is determined by a single gene.

2. Each parent of the cross carries two such genes, one in each of a pair of homologous chromosomes.

3. The gene comes in two versions, one makes seeds round, the other makes them wrinkled. We can refer to the two versions of the gene as R (round) and r (wrinkled). We call different versions of the same gene alleles.

4. The parents of the cross are pure breeding. Therefore they each contain two copies of only one version of the gene. One parent is

RR, signifying two genes for roundness, the other is rr with two genes for wrinkled. The cross can be written as: RR x rr. When an individual has two identical genes for a single character, as have the parents in this cross, it is called homozygous.

5. In the production of gametes only one gene goes into each gamete. This distribution of genes to gametes is called segregation. The RR parent produces gametes of type R and the rr parent gametes of type r. There is no genetic distinction between male and female gametes with respect to these genes. In terms of what happens at meiosis, only one of a pair of homologous chromosomes (and one copy of all the genes it carries) ends up in each of the meiotic products.

6. When gametes fuse they bring together the genes they each carry, and restore the number of genes for a single character to two. One R gamete meets one r gamete to produce an individual of the F1 generation which is Rr. In fact all the F1 generation are Rr. In other words they carry two different versions of the gene for seed shape. The seeds are round because R is dominant over r.

Production of the F2 generation

7. When the F1 plants (Rr) undergo meiosis and the genes segregate into the gametes, half the gametes are going to have the R version of the gene and half the r version. So what happens at fertilisation?

8. The fusion of gametes at fertilisation is a random process. In other words every male gamete has an equal chance of fertilising a given female gamete and every female gamete has an equal chance of being fertilised by any particular male gamete.

In the self fertilisation of F1 plants, therefore, there are four equally possible ways for gametes to fuse:

> R with r (R from the male side)
> r with R (r from the male side)
> R with R
> r with r

The resulting F2 genotypes from these possible fusions of gametes are, respectively:

Rr round
rR round
rr wrinkled
RR round

The first two of these are, effectively, the same. Once the R gene and the r gene come together at fertilisation their origin, whether from the male or the female side, becomes irrelevant.

Because the round character is dominant, all the first three types will be round seeds. Only the last will be wrinkled. That gives an F2 group of three quarters round and one quarter wrinkled, a ratio of three to one.

Summary
To summarise the key features of this interpretation of Mendel's three to one ratio in the F2 of a monohybrid cross:

1. pairs of genes
2. segregation of one of a pair into gametes
3. different versions (alleles) of one gene
4. dominance and recessiveness
5. random fertilisation

The Punnett square
To work out the results of a cross when the parental genotypes are known, it is common to use a Punnett square (named after its originator, Reginald Punnett). For this cross Rr x Rr the Punnett square looks like this:

	R	r
R	RR	Rr
r	Rr	rr

The Punnett square is a device to calculate chance. Across the top is each possible genotype of female gamete. The proportion of each one gives the chance of it being involved in a fertilisation – in this case one in two. Down the side are the possible male gametes. The rest of the table gives the genotypes resulting from every possible fertilisation.

Calculating the chance of each genotype in the progeny
In the example above, there are a total of four combinations of

gametes and only one results in the double recessive genotype rr. The chance of the double recessive wrinkled seeds appearing in the progeny of the cross is, therefore, one in four.

▶ *Key tip* – Avoid the meaningless mistake of repeating one possible type of gamete. For example, a Punnett square for the first monohybrid cross outlined above should be:

	R
r	Rr

or

	r
R	Rr

not:

	R	R
r	Rr	Rr
r	Rr	Rr

as is sometimes drawn.

Remember that the number of genotypes along the top and down the side merely records the chance of each of them occurring. If all the gametes from one or other side of a cross are of one genotype that genotype should only appear once in the Punnett square (= 100% chance).

Genotype and phenotype

These terms cropped up and were defined in Ch. 5. Here it is even more important to understand the difference. Both the seeds which are RR and those which are Rr, have the round phenotype, even though they have different genotypes. Where the two genes are not identical (Rr) the individual is said to be heterozygous (cf. homozygous, page 119). It is impossible to tell whether a round seed is homozygous or heterozygous, except by carrying out a genetic test. But a wrinkled seed must be homozygous because the r allele is recessive.

The test cross

It is easy to distinguish a heterozygous individual from a homozygous one by means of a simple test. The individual in question is crossed with a second individual that shows the character in question in its recessive form. In the case of round and wrinkled seeds, for example, the round seeds would be grown on and the resulting plants crossed with wrinkled seeded plants. There are two possibilities:

(i) homozygous round parent   (ii) heterozygous round parent
   RR x rr   Rr x rr

	R
r	Rr

	R	r
r	Rr	rr

We know the wrinkled seeded plant is homozygous because it shows the recessive character. The only question is whether any wrinkled seeds will be produced in the cross. If they are produced, the other parent must be heterozygous. In fact, as you can see from the second Punnett square, a cross between a double recessive and a heterozygous plant will produce equal numbers of progeny of the two types, a ratio of one to one, round to wrinkled.

A cross designed to test the genotype of one of the parents is known as a test cross.

Dihybrid crosses

Mendel also investigated crosses between parents differing with respect to two characters. In doing so, he discovered the fourth of his principles, namely the principle or law of independent assortment (the first three of Mendel's principles have been mentioned: (i) genes in pairs, (ii) dominance and recessiveness, (iii) segregation).

Take, for example, the cross between a pure breeding plant with round seeds which are yellow in colour and a pure breeding plant with wrinkled, green seeds:

RRYY x rryy (Y = yellow seeded, y = green)

The F1 are all heterozygous round and yellow. RrYy.

This can easily be shown with the relevant Punnett square.

	RY
ry	RrYy

When the F1 plants fertilise themselves what sorts of gametes do they produce? We can work backwards from Mendel's actual results. In the F2 progeny he found:

315 Round Yellow
101 Wrinkled Green
108 Round Green
32 Wrinkled Green

All possible combinations were present in a ratio which was close to 9:3:3:1.

Mendel's conclusion was that when the double heterozygous F2 parent forms gametes it produces all possible types in equal numbers. We can show this in a Punnett square. Notice that there are sixteen divisions of the square, this being the total of $9 + 3 + 3 + 1$. In terms of chance, therefore, the different phenotypes are expected to occur amongst the progeny: nine times out of sixteen, three times out of sixteen, three times out of sixteen and once out of sixteen.

RrYy x RrYy

	RY	Ry	rY	ry
RY	RRYY	RRYy	RrYY	RrYy
Ry	RRYy	RRyy	RrYy	Rryy
rY	RrYY	RrYy	rrYY	rrYy
ry	RrYy	Rryy	rrYy	rryy

Check the outcome of this square and see that it gives a 9:3:3:1 ratio of phenotypes, as Mendel found. To do this, use the rule of dominance. Any genotype with at least one R is round and any with at least one Y is yellow.

Independent assortment

What Mendel meant by independent assortment was the way in which the two factors R (or r) and Y (or y) enter gametes in a non-preferential way. If R, for example, finds its way into one gamete it is just as likely to find Y alongside it as y.

The segregation of pairs of genes into gametes is independent of the segregation of other pairs.

Independent assortment and meiosis

Look again at the diagrams showing the meiotic cell division on page 112. The first division is shown as separating the chromosome pairs in one particular way:

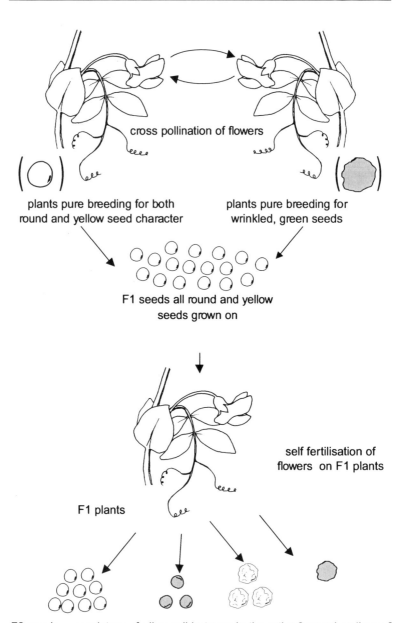

cross pollination of flowers

plants pure breeding for both round and yellow seed character

plants pure breeding for wrinkled, green seeds

F1 seeds all round and yellow seeds grown on

self fertilisation of flowers on F1 plants

F1 plants

F2 seeds – a mixture of all possible types in the ratio: 9 round, yellow : 3 round, green : 3 wrinkled, yellow : 1 wrinkled, green.

Fig. 36. Summary of one of Mendel's dihybrid crosses.

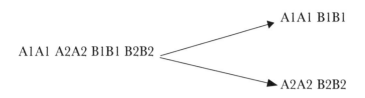

But the alternative way is just as likely:

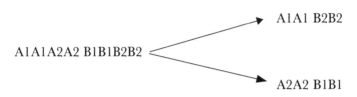

Pure chance governs what actually happens at meiosis. In a large number of meiotic divisions approximately half would segregate the A1 chromosome with the B1 chromosome and the A2 with the B2. The other half would put A1 and B2 together and A2 with B1. This is exactly the same phenomenon as the independent assortment of genes established by Mendel. If the genes that control two separate characters are on different pairs of homologous chromosomes they will assort independently. In peas the gene that controls seed shape and the gene that controls seed colour are, indeed, on different chromosomes.

The classical gene

Mendel referred to unit factors, these being the particulate controllers of the single character differences between his experimental plants. We call them genes. In chapter 3, the gene was defined in molecular terms. We can now be clearer about the classical definition of the gene. A gene is (i) a controller of a single character of an organism, (ii) a unit of inheritance that segregates independently, and (iii) a specific part of a chromosome.

This latter fact will become clearer in the next chapter. Each gene has its own particular place or locus on the chromosomes of a particular homologous pair.

Confusion between gene and allele

There is often needless confusion between the meaning and proper usage of gene and allele.

▶ A gene is defined as above, by what it controls, by how it behaves during inheritance and by where, on a chromosome, it is found.

▶ An allele is a specific version of a particular gene. A gene may have two, three or more allelic forms.

▶ *Key tip* – If the post of prime minister of a country is analogous to gene, then an allele of that gene would be the particular prime minister at any one time. If a new prime minister is appointed (a different allele) the post (the gene) doesn't change though the job might be done in a different way and with different results. You should find this analogy goes as far as necessary in helping you to avoid any confusion in using the terms gene and allele.

Simple exceptions to Mendelian ratios

The occurrence of different phenotypes in the progeny of a cross in ratios that are recognisably Mendelian (3:1 and 9:3:3:1, for example) signifies that Mendel's four principles apply. Other ratios can occur, however, when the results of Mendelian inheritance are modified by an additional factor or when any of the principles of paired alleles – segregation, dominance and independent assortment – break down. Often these ratios are simple and can be recognised as related to expected Mendelian ratios.

Incomplete dominance

Mendel's 3:1 ratio of phenotypes in the F2 of a monohybrid cross depends on complete dominance of one allele over the other. When neither allele is completely dominant the phenotype ratio becomes the same as the genotype ratio, that is 1:2:1.

In snapdragon (Antirrhinum) plants, for example, the allele for red flowers R1 is incompletely dominant in the presence of the allele for white flowers R2. In a cross of the sort:

R1R2 x R1R2 (pink x pink)

	R1	R2
R1	R1R1 (red)	R1R2 (pink)
R2	R1R2 (pink)	R2R2 (white)

The result is a 1:2:1 ratio of genotypes, as expected, and also a 1:2:1 ratio of phenotypes: red:pink:white. The heterozygous plants have pink flowers, not red, because the red colour is only partly expressed.

Codominance

Codominance is a relationship between alleles in which each has a different positive contribution to make to the phenotype of an organism. The situation is similar to the case of incomplete dominance except that incomplete dominance only requires a positive contribution (for example, red flower colour) from one allele.

An example of codominance is the production of two forms of glycoprotein in human red blood cells. The alleles, M and N each produces its own form of the molecule. Heterozygous individuals have both types of glycoprotein.

Genotype	Phenotype
MM	M type glycoprotein only
NN	N type glycoprotein only
MN	both M and N types of glycoprotein

Lethal alleles

In mice an allele Y which gives a mouse a yellow coloured coat, is dominant over the allele y (normal colour). The cross Yy x Yy would be expected to give a 3:1 ratio of phenotypes: yellow:normal. In fact a cross of this sort always gives a 2:1 ratio: yellow:normal. The explanation lies in the finding that the homozygous condition YY is not viable. The Y allele, in the absence of the y allele is a lethal allele. There is still a one in four chance of a fertilisation of the type Y + Y but no living progeny come from such a combination.

Note that, in this case, the Y allele is dominant for its effect on coat colour but recessive for its lethal effect.

Gene interaction

Mendel's dihybrid crosses involved two quite separate genes controlling two distinct characters. It is possible, however, for more than one gene to affect the same character. The genes may be inherited in a Mendelian manner even though simple Mendelian ratios will not be found in the progeny of crosses.

The case where the expression of one gene pair is modified by the action of another pair of genes is called epistasis.

Polygenic inheritance

When a character is controlled by a number of genes each having an additive effect on the phenotype, the effect can be quantitative rather than discontinuous variation. Such traits as height in tobacco plants, height in humans, skin colour in humans, seed size in the broad bean and grain colour in wheat are believed to be controlled by multiple genes or polygenes. There are many more examples.

In polygenic inheritance each single gene pair may still behave according to Mendelian principles, though this will not be clear from the outcome of crosses.

Cytoplasmic inheritance

Only genes found on chromosomes can be inherited in a Mendelian manner. As has been mentioned, the principles of segregation and independent assortment are a direct result of the behaviour of chromosomes at meiosis. All chromosomal genes owe at least some of their inheritance pattern to Mendel's laws.

There are, however, some genes that are found, not on chromosomes, but in the cytoplasm. Both mitochondria and chloroplasts contain small amounts of DNA with active genes.

How are cytoplasmic genes inherited? Because the female egg cell contributes almost all the cytoplasm to the zygote at fertilisation and the male gamete hardly any, they are inherited solely via the female or maternal side.

Mitochondrial and chloroplast genes, therefore, are inherited in exactly the same way as if the mother in a cross had reproduced asexually.

Tutorial

Progress questions

1. What can you recall about chromosomes?

2. How does the behaviour of chromosomes at mitosis and meiosis provide the key to understanding the inheritance of characters in asexual and sexual reproduction respectively?

3. What do the following terms mean?

   (a) haploid and diploid
   (b) homologous chromosomes
   (c) chromatid
   (d) centromere

4. What are the essential differences between mitosis and meiosis?

5. What do the following terms mean?

   (a) gene and allele
   (b) dominant and recessive
   (c) segregation
   (d) monohybrid cross and dihybrid cross
   (e) independent assortment
   (f) homozygous and heterozygous

Seminar discussion

What part does chance play in Mendelian inheritance and at what stages? Is chance predictable or unpredictable?

Practical assignment

To show how chance, allied to Mendel's principles, brings about the anticipated ratios of progeny in simple crosses, carry out the following exercise:

(a) Take two coins. Each represents a parent in a monohybrid cross. Each parent is heterozygous, Hh (H for heads and h for tails).

(b) Toss the first coin once to make a gamete. If it lands heads up, the gamete carries the H allele. If it lands tails up, the gamete is h. Now toss the second coin to make a gamete from the other parent. Put the two gametes together to make the first individual of the next generation.

(c) Keep tossing the coins and recording the genotypes of individuals until a recognisable ratio starts to appear. If you record enough progeny you should see a ratio of 1:2:1 for genotypes or 3:1 for phenotypes, assuming that H is dominant.

You can extend this exercise to dihybrid crosses. You will need two coins for each parent and heads or tails will represent a different pair of alleles in each case. See if you generate a 9:3:3:1 phenotype ratio.

Study tips

1. To recognise Mendelian ratios from the results of crosses always bear in mind that for a monohybrid cross the total comes to 4 (3 + 1 or 1 + 2 + 1) and for a dihybrid cross the total comes to 16 (e.g. 9 + 3 + 3 + 1).

2. Suppose a cross gave 282 of one type of progeny and 101 of a second type. Is this a 3:1 ratio? There is a statistical test for the significance of fit for data of this sort. It is called the chi square test and you should become familiar with it.

3. Don't let Mendelian genetics become a mechanical exercise in the use of symbols, Punnet squares and ratios. Always think about what is involved in the production of each different genotype of progeny, from formation of gametes to final individual.

4. To help you do this, make sure you are absolutely familiar with the genetically significant events of meiosis; namely, replication, homologous pairing, segregation and independent assortment.

Genetics of Higher Organisms II
Linkage, crossing over and chromosome mapping

One minute summary – Genes found on the same chromosome tend to be inherited together and are said to be linked. Linked genes do not follow Mendel's principle of independent assortment and crosses involving linked genes will not give the same Mendelian ratios as crosses involving genes on different chromosome pairs. Linked genes can only be separated by breakage and reunion of chromosomes, an event called 'crossing over'. The likelihood of crossing over occurring between two genes becomes greater the further apart the genes are on the chromosome. This principle can be used to make chromosome maps. Genes that are found on the sex chromosomes are said to be sex linked.

In this chapter we will explore:

▶ the meaning of the terms linkage and sex linkage
▶ how linked genes and sex linked genes behave
▶ how to recognise linkage from the results of crosses
▶ what crossing over is and how it affects the inheritance of linked genes
▶ the importance of the centromere in crossing over
▶ how chromosomes are mapped from the results of crosses

Linkage

Genes on different chromosomes assort independently. This is because, as you saw in the last chapter, the movement of one pair of homologous chromosomes during meiosis is independent of the movement of any other pair. But what if genes are on the same chromosome? Genes that are held together in this way are said to be linked. All the genes that lie on one chromosome are said to form a linkage group.

Behaviour of linked genes

Suppose the gene A with its allele a, and the gene B and allele b are linked on the same chromosomes. There are two ways they can be linked:

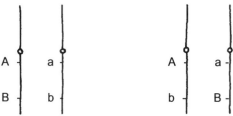

After replication, at metaphase of the first meiotic division, the situation, in each case will be:

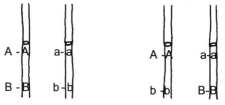

and the four products of meiosis in each case will be:

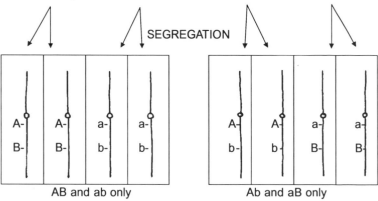

Fig. 37. Segregation of linked genes.

You can see that the genes have not assorted independently and that there are only two, not four, different types of meiotic product in each case. The genes are still linked, just as they were before meiosis. The linked genes have segregated together.

Recognising linkage from the results of crosses

Take a dihybrid back cross of the sort:

Aa Bb   x aa bb

If the genes are not linked (they assort independently), the result of the cross will be:

	AB	Ab	aB	ab
ab	AaBb	Aabb	aaBb	aabb

- a ratio of 1:1:1:1 of the four possible genotypes and phenotypes.
  If the genes are linked A to B and a to b:

	AB	ab
ab	AaBb	aabb

- a ratio of 1:1 of only two types of progeny. Notice that the progeny are **parental types** (the same as the parents of the cross).
  If the genes are linked A to b and a to B:

	Ab	aB
ab	Aabb	aaBb

- again a 1:1 ratio, but the progeny are not like the parents.

▶ *Key tip* – To spot linkage in the results of a cross, look for a restricted number of types of progeny, that is, fewer than you would expect if genes assorted independently. Look, also, for ratios that are characteristic of a monohybrid cross rather than a dihybrid cross.

Try working out the result of the cross: AaBb x AaBb where the genes are linked A to B and a to b. You should find that you get a 1:2:1 genotype ratio and a 3:1 phenotype ratio, the same as for a monohybrid cross.

Crossing over

Genes that are completely linked never segregate into different meiotic products. It is often the case, however, that linked genes can be separated by the process known as crossing over. Crossing over is essentially the same event as the recombination of genes in bacterial and bacteriophage DNA – breakage and reunion of chromosomal DNA, and a consequent exchange of genes between homologous chromosomes.

The crossing over event

Crossing over occurs early on in meiosis, after the replication of DNA but before chromosomes become visible. It involves breakage of two homologous chromatids, at precisely the same site, and the reunion of the strands in a different combination.

(i) breakage

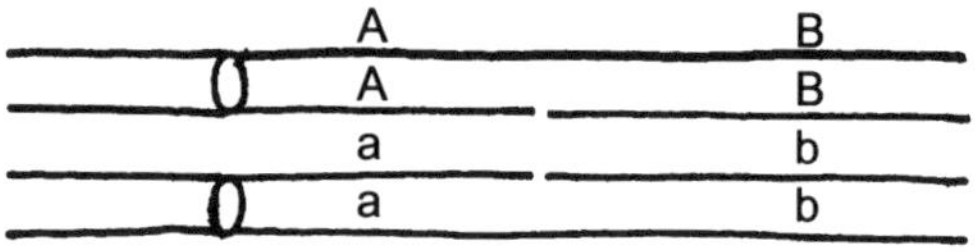

Two homologous chromatids break at precisely the same site:

(ii) reunion

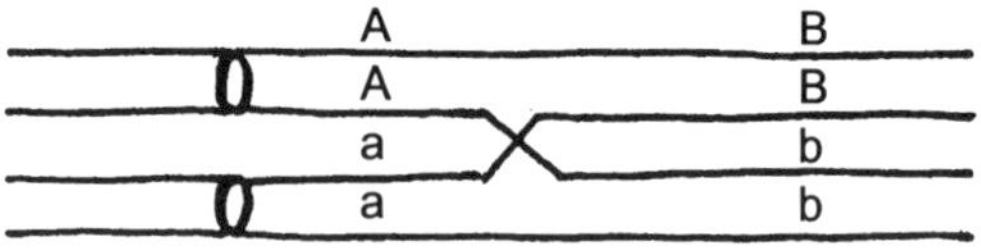

Strands reunite in a different combination.

Fig. 38. Crossing over by breakage and reunion.

Memorise the last of these diagrams. It shows the result of crossing over in four homologous chromatids. It is, in all its variations, one of the most important diagrams in genetics. You will find it a great advantage to be able to reproduce it. But do not make the common mistake of drawing it as:

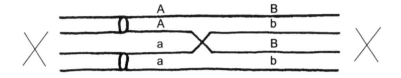

Fig. 39. The wrong way to show the result of crossing over.

To draw crossing over in this way is not only wrong but also shows a lack of understanding of what has happened during meiosis. Breakage and reunion of strands occurs without the repositioning of the strands themselves. Actually, if you examine the incorrect diagram, you will see that no recombination has occurred! The top centromere is still attached to two A alleles and two B alleles, the bottom centromere to two a alleles and two b alleles.

Crossing over and segregation
The effect of crossing over is to change the way in which linked genes segregate at meiosis. Let's follow the fate of the genes A/a and B/b, with and without crossing over between them (fig. 40).

The right hand diagram shows that crossing over has produced two new, recombinant, meiotic products Ab and aB, neither of which is produced in the left hand diagram. Notice, however, that two of the four products of meiosis in the right hand diagram, AB and ab, are the same as those produced in the diagram on the left. Crossing over, like DNA replication, is semiconservative.

The role of the centromere
If you have difficulty working out the genetic consequences of different cross overs, always concentrate on the centromere and what is physically attached to it. At first meiotic division, the two centromeres separate and pull behind them all the genes that are attached to them, in whatever way they have been recombined by crossing over. Remember that crossing over is an actual physical

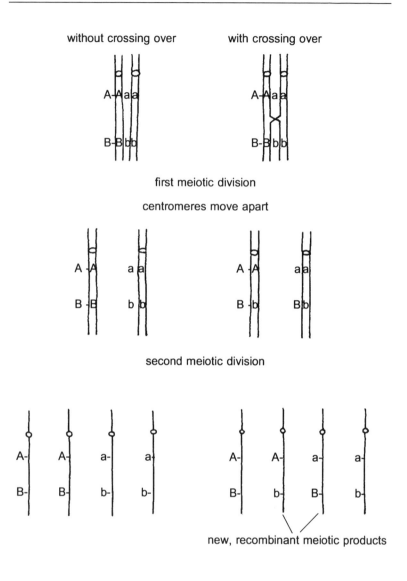

Fig. 40. The effect of crossing over on segregation of two linked pairs of alleles. Note that, when there is no crossing over, both pairs of alleles segregate – A separates from a, and B from b – at the first meoitic division. However, if crossing over occurs between the two genes, segregation of B and b doesn't occur until the second division.

event, like uncoupling the coaches in two trains and swapping them to the opposite engine. The centromere, like the engine of a train, with its coaches in tow, controls which genes go where. Once the genes have been coupled up to a different centromere by crossing over they have no choice but to go where the centromere takes them.

Multiple cross overs

Using the above advice regarding the centromere, you should be able to work out the genotypes of the four meiotic products from the following example. There are three linked genes A/a, B/b and C/c. Just follow the lines and decide which alleles are now attached to each centromere. Remember that the arrangement of genes on the chromatids cannot change after the first meiotic division.

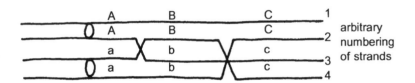

Notes
Strand 1 is not involved in any cross over.
One cross over involves strands 2 and 3.
A second cross over involves strands 2 and 4.

Fig. 41. Multiple cross overs.

A diagram of this sort can only be meaningfully drawn because:

(a) Crossing over occurs after DNA replication, at the 'four strand' stage of meiosis.

(b) All four strands can be involved in cross over events, in any combination.

(c) Each single strand may be involved in none or any other number of cross overs.

(d) Only two strands are involved in any one cross over.

The answer to the problem is: ABC, Abc, aBc and abC.

▶ *Key tip* – In problems of this sort always check that you haven't lost or gained any alleles. There should still be a total of two of each type of allele in the four meiotic products.

Tetrad analysis

In some fungi, and a few algae, it is possible to examine the products of a single meiotic division. This has been a valuable genetic tool. Among other things, it provides evidence that crossing over occurs at the four strand stage of meiosis and can involve any of the four strands.

In fungi of genera such as Sordaria and Neurospora, spores are produced in linear groups of eight, the result of meiosis followed by one mitotic division. Because the mitotic division merely copies each of the four cells produced by meiosis exactly, the groups of eight spores can be considered as if they were a group of four.

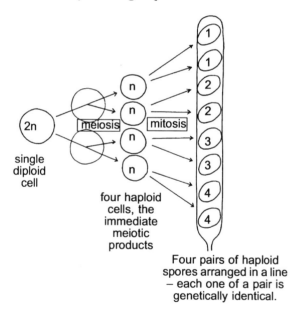

Fig. 42. Spore production in a fungus such as Sordaria.

The individual spores can be removed from their spore case and tested for genetic characters. This is what is meant by tetrad analysis (tetrad referring to the four products of meiosis). In this way the precise events of single meiotic divisions can be determined. The following example demonstrates the principle behind tetrad analysis (fig. 43).

In Sordaria a gene for spore colour has two alleles, b for black spores and w for white spores. The question of dominance does not arise because spores are haploid. When a black spored and a white spored variety of the mould are crossed, a diploid heterozygous stage, bw, is formed. When spores are subsequently produced, half are black and half white.

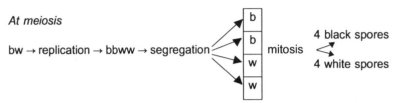

The interesting thing is that there are six possible arrangements of black and white spores in different groups of eight:

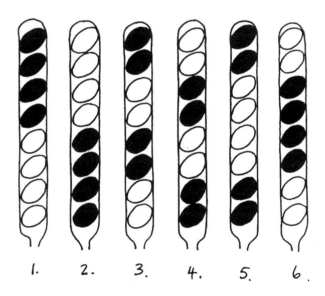

How do these come about? Patterns 1 and 2 are produced when there is no crossing over between the spore colour gene and the centromere of the chromosome on which the gene is found. Which of the two patterns is produced depends on the random segregation of the alleles b and w, whether upwards or downwards.

Fig. 43. Spore colour in sordaria.

e.g. arrangement 1

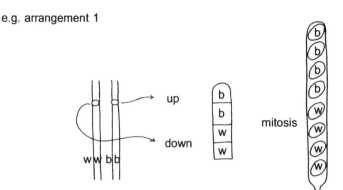

Arrangement 2 happens when segregation is reversed (ww up, bb down).
Note that in both cases 1 and 2 segregation of alleles occurs at the *first* meiotic division.

All other arrangements of black and white spores can only come about when crossing over occurs between the gene and the centromere.

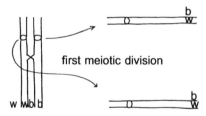

Note that, in this case, different alleles do not segregate at the first meiotic division (b and w are still paired together in the same cell). Segregation must happen at the second meiotic division.

There are four ways for the b and w alleles to segregate.

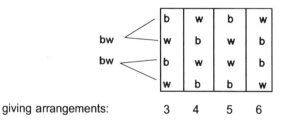

	b	w	b	w
bw	w	b	w	b
bw	b	w	w	b
	w	b	b	w

giving arrangements: 3   4   5   6

You can see that a lot can be deduced about the precise events of meiosis just by looking at the arrangement of different spores.

Fig. 43. (continued)

Genetic crosses involving crossing over

When we considered the cross AaBb x aabb, where A is linked to B and a to b (page 133), we concluded that only two types of progeny could be produced. They would be parental types, AaBb and aabb.

Crossing over can change all that. Crossing over, as was shown in the diagram on page 136, can produce recombinant meiotic products. So the double heterozygous AaBb parent in the above cross can, when crossing over occurs between the two genes, produce all four possible types of gamete. Four types of progeny are, therefore possible from the cross.

The chance of crossing over

The results of a cross usually represent the fusion of a large number of gametes taken at random. Some of these gametes will have come from meiotic divisions in which crossing over between genetic markers occurred, and some from divisions where it did not.

Example

A heterozygous grey bodied (Bb) and normal winged (Vv) female fruit fly (crossing over only occurs in female flies) was crossed with a black bodied (bb), vestigeal winged (vv) male.

BbVv x bbvv (a standard test cross)

Progeny:

grey, normal	(BbVv)	126
black, vestigeal	(bbvv)	124
grey, vestigeal	(Bbvv)	24
black normal	(bbVv)	26

Conclusions

(a) The results do not show a 1:1:1:1 ratio – linkage of genes is indicated.

(b) Parental types predominate in the progeny – the genes are linked in the manner B to V, b to v.

(c) Recombinant progeny have appeared – crossing over must have occurred.

(d) The recombinants represent $\dfrac{24 + 26}{24 + 26 + 124 + 126}$ x 100%:

= 25% of the progeny.

(e) In the production of gametes by meiosis crossing over between the genes must have occurred in roughly half of all cases (25% x 2).

Think about this last conclusion. Remember, crossing over is semi conservative. Only half the haploid products of a single cross over are recombinant. This is why the frequency of recombinant progeny is half the frequency with which crossing over occurs during meiosis.

Chromosome mapping

The principle of mapping the genes on a chromosome is very simple. The likelihood of a cross over happening between two genes is related to the distance between them. Two genes that are very close together on the chromosome are only rarely going to be separated by crossing over. Conversely, genes that are far apart are extremely likely to be separated in this way.

When genes are separated by crossing over the result is recombinant gametes and, in due course, recombinant progeny. The frequency with which recombinants appear in the progeny of a cross involving linked genes is used as a direct measure of the map distance between the genes.

▶ *Key point* – 1% recombinant frequency = one unit of map distance.

In the previous example the frequency of recombinant progeny was 25%. Therefore the genes for body colour and wing type are 25 map units apart on the chromosome.

Establishing the order of genes

Recombinant frequency gives the relative distance apart of genes on a chromosome map, but how is the order or sequence of the genes determined? If enough recombinant frequencies are known, this information alone will do the job.

For example, on the same chromosome as the gene for black body and the gene for vestigeal wings in Drosophila is a gene for purple eyes (P/p). If we have three recombinant frequencies we can not only put the genes at the proper distances apart, but also put them in the right order.

Recombinant frequencies (map distances)
B/b – V/v 25% (see example above)
B/b – P/p 11% (from the cross BbPp x bbpp)
P/p – V/v 14% (from the cross PpVv x ppvv)
This information gives a map:

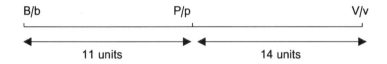

Fig. 44. Genetic map for three linked gene loci.

Note that two map distances alone would be insufficient to establish the order of the genes.

Genetic map and physical chromosome map
Is a genetic map constructed from recombinant frequencies an accurate representation of the actual physical map of the chromosome?

In one respect it is. The order of genes established from the results of crosses is the actual order of genes on the chromosome. As far as the distances apart of the genes is concerned, however, the genetic map is not accurate.

(a) A genetic map only gives relative distances based on statistical comparisons. These distances do not have any length units attached to them.

(b) The chance of crossing over occurring is not the same in all parts of the chromosome, as is assumed for the purposes of a genetic map. One map unit (1% recombinants) is, therefore, not a constant physical distance.

(c) In some parts of a chromosome crossing over is inhibited altogether. All the genes on such regions of the chromosome, even though the region has a physical length, would appear to map at exactly the same place with no map units between them.

Effect of two cross overs

Genes that are very far apart on a chromosome can be recombined by one cross over and then put back in their original combination by a second cross over. This causes map distances between genes to be underestimated and, for this reason, genetic maps are usually made by adding lots of small distances together rather than dealing with genes that are widely separated.

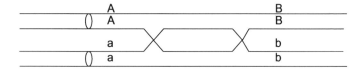

meiotic products: AB, AB, ab, ab – no observed recombination

Fig. 45. The effect of two cross overs.

Sex linkage

Chromosomal sex determining mechanisms

In a number of higher organisms, mammals, birds and insects, for example, maleness and femaleness is determined by a chromosomal mechanism. The most common type is the XY sex determining mechanism.

X and Y chromosomes

The two partners of one particular pair of chromosomes, the sex chromosomes, unlike the individual chromosomes making up the other pairs in a full set, are not entirely homologous. In man, for example, the X chromosome is considerably longer than the Y chromosome. The X and Y chromosomes do pair at meiosis but only over a very small homologous part of each.

In man, and all other mammals, females have two X chromosomes plus the autosomes and males have the same autosomes plus one X and one Y chromosome. Autosomes is the name given to chromosomes which are not sex chromosomes. The chromosomal distinction, XX or XY is sufficient by itself to determine which sex an individual will be.

Segregation of sex chromosomes

At meiosis in XX females only one type of gamete can be produced with respect to the sex chromosomes. All eggs will contain one X chromosome.

In males half the sperm will contain an X chromosome and half will contain a Y chromosome. In effect the segregation of the X and Y chromosomes follows the same pattern as the segregation of single genes according to Mendel's principles.

Because XY individuals produce two types of gamete they are referred to as the heterogametic sex. This is not always the male sex. In birds it is the female that is heterogametic.

A cross between two individuals can be represented by a Punnett square. Only one cross is possible because there is no such thing as a cross between two males or between two females!

	X
X	XX
Y	XY

The result of any cross is parental types (male and female) in equal proportions. This is the simple mechanism by which males and females are maintained in equal numbers in a population.

Genes on the sex chromosomes

Any genes that occur on the sex chromosome are said to be sex linked. There are three possible ways in which a gene may be sex linked:

1. It may be linked permanently to the Y chromosome and have no counterpart on the X chromosome. This is Y linkage.

2. It may be linked permanently to the X chromosome and have no counterpart on the Y chromosome. This is X linkage.

3. It may be found on both X and Y chromosomes and be capable of being exchanged by crossing over. This is referred to as incomplete sex linkage.

These three types of sex linkage are possible because:

(a) There is, in some animals, a small part of the Y chromosome which carries some genes and has no partner in the X chromosome.

(b) There is a considerable part of the X chromosome which has no homologue in the Y chromosome.

(c) Small parts of the X and Y chromosomes are homologous and carry genes in pairs like other chromosomes.

Behaviour of sex linked genes

Y linked genes

The behaviour of Y linked genes is totally predictable. They pass directly from the heterogametic parent to the heterogametic progeny in every case and can be inherited in no other way.

An individual can only have one allele of a Y linked gene since there is no partner on the X chromosome. The problem of dominance and recessiveness, therefore, does not arise.

Not many Y linked genes are known. Hairy ears in men is said to be a Y linked trait. Its inheritance is exclusively from father to son.

X linked genes

The XX sex will have two copies of an X linked gene and can, therefore, be homozygous or heterozygous with respect to the alleles that are present. The XY sex, on the other hand, only has one allele, there being no place for one on the Y chromosome. This has implications for the expression of dominant and recessive alleles. Consider, for example, a pair of alleles of an X linked gene: A and a:

Sex	possible genotype	phenotype
XX	AA	A
XX	aa	a
XX	Aa	A
XY	AY	A
XY	aY	a

Note: AY and aY signify the presence of a single allele on the X chromosome, accompanied by a Y chromosome from which the gene is missing.

Pattern of inheritance of an X linked gene

What characterises X linked inheritance is that the X chromosome in an XY individual, along with the genes it carries, must have come from its XX parent and cannot have come from its XY parent. There is a tendency, therefore, for inheritance to be biased in the direction: homogametic – heterogametic sex. In human terms, sons inherit their X linked genes from their mothers while daughters get an equal contribution of X linked genes from each parent.

Colour blindness: an example of an X linked character

One human X linked character is a form of colour blindness. The dominant allele, A, gives normal vision, the recessive allele, a, is responsible for colour blindness. Refer back to the table of genotypes and phenotypes to see the possibilities for colour blindness in males and females.

Suppose a heterozygous woman has children by a normally sighted man:

Aa x AY

	A	a
A	AA	Aa
Y	AY	aY

None of the daughters will be colour blind, but they will stand a one in two chance of carrying the recessive allele. Sons will have an equal chance of being colour blind or normal sighted.

If a homozygous normal-sighted woman has children by a colour blind man:

AA x aY

	A
a	Aa
Y	AY

None of the children will be colour blind but all the daughters will carry the recessive alllele.

Incompletely sex linked genes

Incomplete sex linkage is characterised by behaviour similar to that of genes linked on autosomes. There is a strong tendency for inheritance to be linked to the sex of progeny in the same way that parental types can predominate in the progeny of crosses involving autosomally linked genes. But the linkage can change as a result of crossing over and is not, therefore, permanently fixed, as in the case of X and Y linkage.

Tutorial

Progress questions

1. Define the terms recombination, recombinant and recombinant frequency.

2. In tomatoes, tall is dominant to dwarf and round fruit shape is dominant to pear shape. A tall tomato plant bearing round fruits was crossed with a dwarf plant with pear shaped fruits. The result was:

   81 tall, round fruited plants
   79 dwarf, pear fruited
   22 tall, pear fruited
   17 dwarf round fruited

   Are the genes for height and fruit shape linked? If so how were the alleles linked in the parents of the cross, and what is the map distance between them on the chromosome?

3. What is the maximum possible map distance between two genes on a chromosome derivable from the results of a single cross? Is this the same as the maximum actual map distance between two genes? If not why is there a difference?

4. What is sex linkage? What are the three types of sex linkage?

5. How can tetrad analysis tell us more about the events of meiosis than an analysis of the progeny of crosses involving higher plants or animals?

Seminar discussion

1. Why don't linked genes assort independently?

2. What is crossing over and how does it happen?

Practical assignments

1. It may help your appreciation of the events of crossing over to translate the diagrams into models using string, pipe cleaners or insulated electric wiring. Models demonstrate much better than diagrams the physical exchange of material, and the connection of parts of chromosomes to the centromere.

2. Find out more about the molecular events of the breakage and reunion model of crossing over. Is this mechanism the same in higher plants and animals as in bacteria and viruses?

Study tips

1. Learn the diagram on page 134, showing the result of crossing over. Make sure you understand what it represents, how it has come about and its genetic consequences. It really is the key to an understanding of the content of this chapter.

2. Extend your understanding of the principles behind this diagram by involving more than two genes and more than one cross over.

8

Origin and Maintenance of Variation

One minute summary – All genetic variation ultimately owes its origin to the mutation of pre-existing genes. The number of gene loci for an organism, and the consequent potential for variation, can be increased through addition of parts of chromosomes, whole chromosomes, or whole chromosome sets. The mixing of genes during the sexual process produces additional genetic variety from existing variation. Outbreeding mechanisms help to maximise and maintain this sexually generated variation. The use of artificial mutagens suggests how spontaneous mutation might happen and has been of value in genetic research. In this chapter you will learn about:

▶ the conservative nature of inheritance
▶ how sexual reproduction maintains genetic variation
▶ the importance of heterozygosity and how it is maintained by outbreeding
▶ how gene mutation and chromosomal changes are the ultimate sources of genetic change
▶ how different mutagens affect DNA
▶ what changes can occur to chromosomes and chromosome sets and the significance of these changes

The conservative nature of inheritance

As was pointed out in chapter 1, the genetic mechanism is a conservative one. DNA replication is usually an extremely accurate event, resulting in genotypes being perpetuated without change. Even when reproduction is linked to a sexual process the result is not greatly different. Any variation generated is insignificant compared to the similarities between one generation and its progeny.

Variation through the sexual process

You will also recall from chapter 1, and later chapters, that sex produces variation by four means:

1. fusion of genetic material from two different sources
2. segregation of alleles
3. independent assortment of chromosomes
4. recombination by breakage and reunion of homologous DNA.

These things together are rather like the shuffling and dealing of a pack of playing cards to give different hands. They ensure that sexually produced progeny are all different.

▶ *Key point* – It is important to realise, however, that sexual processes can only rearrange pre-existing variation and produce new combinations of old genes. They cannot manufacture genetic change *de novo*.

Gene mutation

▶ *Key point* – Ultimately the only source of changes in the genetic material is mutation. Gene mutation is a spontaneous or induced alteration of the molecular form of a gene that results in the creation a new version of the gene.

The molecular basis of gene mutation

A gene is a sequence of nucleotides and nucleotide base pairs in a DNA molecule. Installed in the sequence is the genetic code specifying the particular protein that the gene is responsible for making. Any change to the code constitutes a mutation. It can cause the gene to initiate the synthesis of an altered protein or, in some cases, no protein at all.

Examples of changes to DNA structure caused by mutation
1. deletion of one or more nucleotide bases
2. addition of nucleotide bases
3. substitution of one base by another
4. other chemical damage, such as the creation of thymine dimers by ultra violet light (see page 41).

Mutagens

Any substance or influence that causes mutation of the genetic material is referred to as a mutagen. Listed below are some examples of mutagens, the specific alteration they make to the DNA molecule and the subsequent consequence of this change. Mutagens have been useful tools in genetic research as they can cause specific and predictable changes to DNA structure.

1. *Base analogues*
These are related compounds to the normal four bases of DNA. They can be incorporated into a DNA molecule in place of the correct base during synthesis. The base analogue 5-bromouracil, for example, replaces thymine. Subsequently 5-bromouracil can base-pair with guanine rather than adenine, giving rise to a mutant gene in which adenine bases are replaced by guanine bases.

2. *Alkylating agents*
By altering the chemical structure of the bases of DNA this group of mutagens change the base pairing rules and, like base analogues, cause base substitutions. The mustard gases, sometimes used as chemical weapons, are alkylating agents.

Substitution of bases in the DNA code, whether induced by alkylating agents, base analogues or by other means, inevitably, alters the code written into a gene. The consequence of a code change involving a single base substitution on the expression of a gene can be negligible, minor or catastrophic. Take the DNA code ACA, for example. The final A of the triplet could be changed by mutation, to G, C or T, with the following consequences:

Change	Consequence
ACA to ACG	No effect, since both triplet codes specify the amino acid cysteine.
ACA to ACC	Substitution of the amino acid tryptophan in place of cysteine in the gene product. This might be very damaging to the function of the protein in question, or it might have little or no effect.
ACA to ACT	Termination of the gene product at this point in the chain, since ACT is a terminator triplet. The mutation would most likely result in no functional gene product being formed.

3. *Acridine dyes*

Acridine dyes cause deletions or additions of bases in the DNA. These, in turn, cause misreading of the genetic code and are known as frame shift mutations for that reason. The code ACAACAACAACA, for example, is read as four similar triplets, ACA, specifying a piece of a protein consisting of four consecutive cysteine molecules. If the first A is deleted the code becomes three CAA (valine) triplets and an incomplete triplet CA.

If a base is added, say an A to the beginning, the code now consists of repeating AAC triplets, translating into four leucine molecules.

Frame shift mutations make complete nonsense of the code in a gene and almost invariably, therefore, result in no useful gene product.

4. *Radiation*

Ultra violet light, X rays and other ionising radiations, such as gamma rays, are all mutagenic. Their effects could be described as more randomly damaging than those of the chemical mutagens already mentioned. The degree of damage caused by radiation can vary from point mutations (changes to single bases of the DNA) to gross changes to chromosomes.

Spontaneous mutation

The causes of spontaneous mutations are unclear, though the presence of natural mutagens in the environment is likely to play a part. Changes to DNA occur by accident, most often during replication, at a very small but significant rate in all organisms. Although most such mutations are unlikely to improve the function of a gene, and may be deleterious, they can be maintained in a population for three reasons:

1. They may alter a gene product without rendering it non functional. Amino acid substitutions in many enzymes, for example, need not destroy the catalytic effectiveness of the enzyme. An enzyme controlling a single unique biochemical reaction can often come in several varieties with slightly different amino acid compositions.

2. They may cause a change in phenotype which is not immediately harmful to the organism's survival and reproduction.

3. The mutated gene can shelter in the presence of a dominant normal allele in the heterozygous state in diploid organisms.

Although it is true to say that most mutations are either neutral or disadvantageous, the longer they can be maintained in a population the longer selective forces have to act on them and to discover any value in them. Modern analytical techniques suggest that there is a huge amount of hidden variation in populations in the form of slightly different forms of the same proteins. This variability may not be detectable except at the biochemical level but it represents a bank of potential material for natural selection to act on in the future.

Spontaneous mutation rates vary from organism to organism and from gene to gene but, at their highest, are of the order of one mutation in every 10,000 cell divisions. In humans the rate is no more than one in every 100,000 cell divisions.

Chromosomal changes

Mutations can not only cause changes to individual genes but, on a grander scale, can alter whole chromosomes or sets of chromosomes. Any alteration of chromosomes is also bound to affect the genes on the chromosomes. It may lead to a rearrangement, a loss or a duplication of genes. Any of these changes, in turn, can alter the expression of the genes or give rise to completely new characteristics, even if the structure of the genes themselves has not been affected.

Changes to the number of chromosomes
1. *Aneuploidy*
The chromosome complement of a cell or an organism can be increased by the addition of one or more whole chromosomes. Alternatively, one or more chromosomes can be lost from a set. In both cases the result is a chromosome number which is not an exact multiple of the haploid set. The situation is called aneuploidy. A good example of aneuploidy is Down's syndrome in humans. Down's syndrome is caused by the presence of one extra chromosome, giving a total of forty seven rather than the normal forty six.

Aneuploidy can result from a failure of a pair of chromosomes to separate properly at meiosis, an accident known as non-disjunction. Non-disjunction will result in gametes with either one extra chromosome or with one missing.

2. *Polyploidy*
Chromosome numbers can be increased to the tune of one or more complete haploid sets on top of the normal diploid complement. Three complete sets gives a triploid cell or organism, four gives a tetraploid and so on. Polyploidy is a general term for multiple chromosome sets.

Extra chromosomes at meiosis

If an organism has extra chromosomes, whether because of aneuploidy or polyploidy, it will run into problems at meiosis. Homologous pairing of chromosomes, which is so vital a feature of meiotic division, requires no more than two chromosomes. Only two can pair together in any one part. When three or more homologous chromosomes are trying to pair, the process is disrupted and rarely successful. Consequently, additions to the normal diploid number of chromosomes often results in sterility.

Structural changes

Chromosomes can break or be broken. Some mutagens, ionising radiations, for example, are known to damage chromosomes in this way. But broken chromosomes can also reunite, not necessarily in the same way as before. A number of structural changes to chromosomes, therefore, can occur (see fig. 46).

The importance of chromosomal changes

Chromosomal mutations do not necessarily involve gene mutation. And since it is genes that control the characters of an organism you might think that chromosomal alterations without gene mutations are not genetically relevant. You would be wrong.

1. *Dosage effects*
The normal complement for any one gene is two alleles. The addition of one or the loss of one can have profound effects on the phenotype, even if nothing new is added or nothing that is apparently essential is taken away. The genotype A and the genotype AAA are not necessarily going to produce the same phenotype as the more normal genotype AA, even though the allele A may be dominant.

Addition of chromosomes to a full human set can have very damaging effects, despite the fact that the genes they carry exist quite harmlessly in a double dose in all normal individuals. We have seen how this is the case, for example, in Down's syndrome.

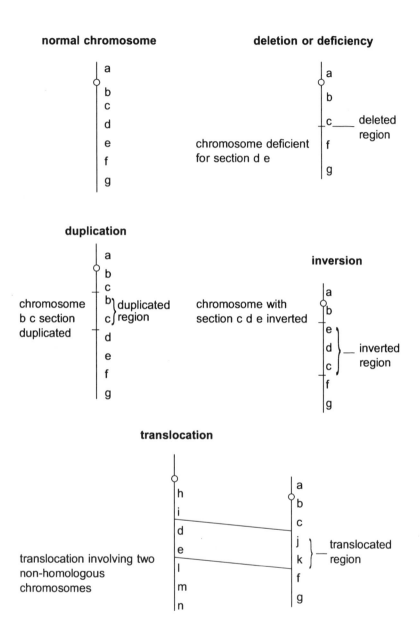

Fig. 46. Types of structural changes to chromosomes.

2. *Position effects*

Any chromosomal mutation that involves an alteration to the position of genes in relation to each other (inversions and translocations, for example) can also alter the expression of the genes. As with dosage effects, it is not necessary for a gene itself to mutate for a phenotypic change to result. The change in position of the gene in relation to other genes is sufficient.

3. *Increase in the number of gene loci*

The evolution of increasing complexity of organisms demands more and more genes. The only way genes (as opposed to alleles) can increase in number is by more gene loci being created. The only way new gene loci can be created is by addition of whole chromosomes or parts of chromosomes to the normal complement. Although the genes carried by these duplications will initially be copies of existing genes, time, mutation, recombination and selection will enable them to become completely different genes.

4. *Repression of crossing over*

It is often advantageous for a group of genes to remain permanently linked because they collectively control some aspect of phenotype and it would be a bad thing for them to recombine with other alleles. If crossing over is repressed in the region of the chromosome carrying these genes, there can be no recombination. Chromosomal changes can contribute to the repression of crossing over.

Deletion of part of a chromosome, for example, will completely rule out crossing over as there will be no homologous partner in that region. This is the basis of the XY sex determining mechanism. The X chromosome carries a number of genes that can never recombine with alleles on the Y chromosome because the equivalent part of the Y chromosome has been deleted.

Inversions and translocations can also lead to a suppression of crossing over. Crossing over within an inverted section of a pair of chromosomes, for example, destroys the physical structure of the chromosomes, creating strong selective pressure for crossing over not to occur in that region (see fig. 47).

A pair of homologous chromosomes, one of which has section a–f inverted.

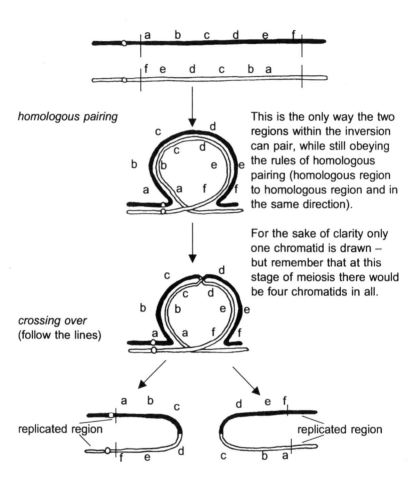

homologous pairing

This is the only way the two regions within the inversion can pair, while still obeying the rules of homologous pairing (homologous region to homologous region and in the same direction).

For the sake of clarity only one chromatid is drawn – but remember that at this stage of meiosis there would be four chromatids in all.

crossing over
(follow the lines)

replicated region

replicated region

Crossing over has resulted in an exchange of genes, as expected, but has also created one chromosome with two centromeres and one with none. You will also see that the parts of the chromosome outside the inversion are now replicated in the two chromosomes. Clearly, this is not a viable situation!

Fig. 47. Crossing over within an inversion.

Maintenance of variation

Homozygosity v heterozygosity

With respect to a single gene locus there is as much variability in one heterozygous individual (Aa) as there is in two homozygous individuals (AA and aa). In populations, therefore, the heterozygous condition is to be preferred for the purposes of maintaining and generating variation.

Heterozygosity brings its own advantages as well. It enables recessive genes to remain in a population even though they are not sustainable in a homozygous state. It also leads to 'hybrid vigour', otherwise known as heterosis. This phenomenon is the basis of the vigour shown by F1 varieties of cultivated plants. Conversely, organisms that are largely homozygous can lack vigour and be at a selective disadvantage.

Inbreeding and outbreeding

Inbreeding is sexual reproduction where the male and female parents are related. In outbreeding there is no relationship between the male and female side in crosses. Inbreeding tends to decrease heterozygosity and outbreeding to sustain it.

Self fertilisation

The most extreme form of inbreeding is self fertilisation. It is easy to see how it increases homozygosity for a gene locus, at the expense of heterozygosity. Refer back to Mendel's monohybrid cross and the production of the F2 generation (chapter 6).

	A	a
A	AA	Aa
a	Aa	aa

Half the progeny of the self fertilisation of the F1, heterozygous parent are homozygous. These progeny, if they continue to self fertilise, will produce more homozygotes. The heterozygous F2 progeny will again give half homozygotes and half heterozygotes. Over many generations of self fertilisation, therefore, the heterozygous state will, effectively, be completely eroded.

Cross fertilisation

Cross fertilisation sustains heterozygosity because most of the possible types of cross involving two alleles of one gene will generate at least half heterozygous progeny:

Type of cross (male x female) *(all possible ones are here)*	*Proportion of heterozygotes* *in progeny*
Aa x Aa	50%
Aa x AA	50%
AA x Aa	50%
Aa x aa	50%
aa x Aa	50%
AA x aa	100%
aa x AA	100%
AA x AA	0%
aa x aa	0%

Outbreeding mechanisms

Selection has favoured outbreeding, and mechanisms that promote it, largely because of the advantages of maintaining heterozygosity. The most obvious of these mechanisms is sexual dimorphism. Related males and females may be able to breed together but they cannot fertilise themselves.

Even when animal or plant species are hermaphrodite there are often devices favouring outbreeding. In many cases self fertilisation is impossible, even though individuals are equipped with both male and female sex organs.

The tendency for living things to mix genes through a combination of a sexual process and outbreeding is of fundamental importance. It is how variety is maintained. The importance of variety to the survival and evolution of animal and plant species was outlined in chapter 1.

Tutorial

Progress questions

1. This question was asked before, at the end of chapter 1, but it's so important that it's worth asking again: how does sex produce variation?

2. What types of mutagen are there? What specific effects do they have on DNA?

3. Why are most mutations either of no obvious value to an organism, or positively deleterious?

4. A part of a DNA molecule has the base sequence: TAC ACC CAC

   (a) What three amino acids does this piece of DNA code for?

   (b) What would be the effect of mutations which substituted a thymine (T) base for (i) the first C in the sequence, (ii) the second C, (iii) the last C?
   (c) and what if a mutation deleted the T at the start of the sequence?

5. What changes can occur to the number and structure of chromosomes? How can these changes affect gene expression?

Seminar discussion
Most organisms have mechanisms that encourage outbreeding. Why is this?

Practical assignments
1. Discover mechanisms which prevent inbreeding in flowering plants.

2. Find out more about chromosomal abnormalities in humans and their resulting syndromes.

Study tips
Be careful about the precise meaning of words when writing about this topic. For example variation and variability mean slightly different things. One describes what already exists, the other what is possible in the future. Mutation and mutant are other words which have to be used carefully.

Population Genetics

One minute summary – Population genetics studies the sum total of all the genes of all the individuals in a population. These genes, together, make up the gene pool for the population. As long as breeding is exclusively within the population, every newly produced individual receives its genes from the existing gene pool. The frequency of different alleles in the gene pool will only change as a result of influences such as selection, mutation, genetic drift and migration. Changes to the composition of a gene pool represent the first step in evolution. In this chapter you will learn:

▶ the meaning of the term population, from a genetic point of view
▶ the importance of the gene pool
▶ how to calculate gene frequencies from observed phenotype frequencies using the Hardy-Weinberg equation
▶ the conditions which apply to a population in Hardy-Weinberg equilibrium
▶ factors that can change the frequency of genes in populations

Genes and populations

What is a population?

A population is a group of individuals of the same species within which mating is occurring or can occur. Implied in this definition is the idea that any individual in the population is a potential mate for any other individual in the same population, or, if the sexes are separate, for any individual of the opposite sex.

From the genetic point of view this is the essential feature of a population. The genes that belong to each individual must be part of a much larger shared bank of genes that are passed around within the population during sexual reproduction. This can only be the case if there are no barriers to the exchange of genes between individuals.

Genes in a population

Each individual in a population has its own genes. With some exceptions they are the same genes as those of every other individual in the group, but not, of course, necessarily the same alleles. All the genes, in their different allelic forms, of all the individuals in a population make up what is known as the gene pool.

The importance of the gene pool is that, provided mating occurs only within a population, the genes and alleles inherited by all the new progeny of the population must come from that gene pool. The gene pool is both a source of variation for the new individuals in a population and a limit to the variation they can show.

Dynamics of genes in populations

Gene pools can change with time. Ultimately, when the time periods being considered become long enough, population genetics turns into the study of evolution.

To understand population genetics we should think of populations, not as collections of genetically distinct individuals, but as single genetic units. The population is often referred to as the 'unit of evolution'. Only when genes move within the population, from individuals of one generation to individuals of the next, and when various factors cause changes to the gene pool, does the study of population genetics take on its full significance. It is a dynamic study. It deals with the movement of genes and with shifts in their relative frequencies.

Gene frequencies

A gene pool comprises all the alleles belonging to all the individuals in a population. Not all alleles of a single gene will be equally common. The measure of how common a particular allele is in a population is known as its gene frequency.

$$\text{Gene frequency} = \frac{\text{Number of copies of that particular allele in the gene pool}}{\text{Total number of all alleles of that gene in the gene pool}}$$

The significance of gene frequencies

Gene frequencies represent the status of alleles in populations and, to some extent, describe the population itself. Populations that differ in character will show different gene frequencies for particular alleles. For example alleles for blue eyes will be more frequent in human populations in Northern Europe than in Mediterranean regions.

Gene frequency and phenotype

Gene frequency must not be confused with phenotype frequency. Dominance and recessiveness of alleles distort their apparent frequency in populations. Recessive alleles are more frequent in populations than the appearance of their phenotype would suggest. Dominant alleles are less frequent. It is quite common for a recessive allele to be more frequent in a population than its dominant counterpart even though the dominant phenotype is the more common one. This is because heterozygous individuals exhibit the dominant phenotype but carry both types of allele.

Changes in gene frequency

A change in the frequency of an allele in a population represents a change in the population itself. It is through changes in gene frequencies that populations evolve. The causes of changes in gene frequencies will be considered later in this chapter.

The Hardy-Weinberg equilibrium

The calculation of gene frequencies

Provided a number of conditions apply, allele frequencies and genotype frequencies remain constant in a population, once a state of genetic equilibrium has been reached. This is the basis of the Hardy-Weinberg equilibrium principle. The principle allows gene frequencies to be calculated for an equilibrium population from observed phenotype frequencies. The principle is named after its originators, G. H. Hardy, an English mathematician and W. Weinberg, a German physician. It is one of the fundamental concepts of population genetics.

Genetic equilibrium

Genetic equilibrium is reached when the frequencies of genotypes in a population remain constant. Genotype frequencies at equilibrium are determined solely by the initial allele frequencies provided:

1. the population is large
2. the members of the population interbreed randomly
3. there is no selection acting in favour of individuals of any particular genotype
4. there is no migration either into or out of the population and no significant effect of mutation or random change in gene frequency.

▶ *Key point* – If these four conditions apply, a Hardy-Weinberg equilibrium will be reached.

Hardy-Weinberg formulae

The rules of Mendelian inheritance and the rules of chance allow mathematical formulae to be calculated which relate allele frequencies in a population to genotype frequencies. These formulae only apply, however, to populations which satisfy the Hardy-Weinberg equilibrium principle requirements.

Where there are only two alleles of a gene in a population (A and a), the relationship between genotypes and allele frequencies is given by:

$$AA : Aa : aa = p^2 : 2pq : q^2$$

where p = gene frequency of the allele A, and q = gene frequency of the allele a. The gene frequencies, p and q are expressed as fractions. If, for example the alleles A and a are equally frequent then $p = 0.5$ and $q = 0.5$.

You should note that because, in this case, there are only two alleles:

$$p + q = 1$$

An example of a Hardy-Weinberg calculation

Some people can taste the chemical substance phenylthiocarbamide (PTC) and some people cannot. The difference is controlled by a single gene with two alleles. The dominant allele, T, is for tasting, the recessive, t, for non-tasting.

In a typical group of 228 people, 160 were found to be tasters, and 68 non-tasters. Tasters may be either homozygous (TT) or heterozygous (Tt). Non-tasters, on the other hand, can only be homozygous (tt).

Since: TT : Tt : tt $= p^2 : 2pq : q^2$

q^2 (tt) $= \dfrac{68}{228} = 0.30$

$\therefore$ q (gene frequency for t allele) $= \sqrt{0.30} = 0.55$

$p + q = 1$

$\therefore$ p (gene frequency for T allele) $= 0.45$

The relative frequencies of homozygous and heterozygous tasters in the population sample are given by p^2 and $2pq$. They work out as 0.20 and 0.50, respectively.

In this example, the recessive allele is the more frequent in the gene pool (0.55 : 0.45), even though the dominant character is more frequently shown in people's phenotypes. The terms 'dominant' and 'recessive' have absolutely no significance in relation to the relative frequencies of alleles in populations.

Another point to note is that provided the population in the above example continues to fit the requirements of the Hardy-Weinberg equilibrium – the gene frequencies for T and t will remain the same from generation to generation, as will the genotype frequencies.

Hardy-Weinberg and multiple alleles
When there are more than two alleles of a single gene in a population the Hardy-Weinberg principle can still apply. The mathematics just becomes a bit more complex.

Human blood groups, for example, are determined by three alleles: A, AB and a. Each allele has its own frequency, p, q and r, respectively:

$$p + q + r = 1$$

If the proportions of the different blood types in a sample of a human population are recorded (column 2 in fig. 48) the following set of data can be calculated:

Blood type phenotype	Number in sample /173	Phenotype frequencies in sample	Genotypes	Frequencies expressed in terms of allelic frequencies
O	78	0.4509	aa	r^2
A	71	0.4104	AA & Aa	$p^2 + 2pr$
B	17	0.0983	ABAB & ABa	$q^2 + 2qr$
AB	7	0.0405	AAB	$2pq$

From the data the allelic frequencies can be calculated:

p   (for allele A)   = 0.26

q   (for allele AB) = 0.07

r   (for allele a)   = 0.67

Fig. 48. Phenotype, genotype and allelic frequencies for human blood types from a population sample of 173 people.

Factors changing gene frequencies in populations

The main factors changing gene frequencies in populations are:

1. selection
2. mutation
3. migration
4. genetic drift
5. meiotic drive.

Selection

Any tendency for one particular allele to be more successful than others at reproducing itself in a population is counted as selection. The result will be that, at each successive generation, as long as selective advantage continues, the relative frequency of that allele will increase at the expense of other alleles.

Selective advantage can come for several reasons:

1. An allele might directly affect the reproductive success of any individual carrying it.

2. It might protect the individual against predators or diseases.
3. It might make it more efficient in its nutrition.

Whatever the nature of the advantage, it will only count as selection in the genetic sense if it leads to a shift in the proportion of alleles in the gene pool. Selection against an allele counts as positive selection with respect to alternative alleles.

Selection acts on phenotypes

The effect of selection on gene frequencies varies depending on whether, and to what extent, the allele that is favoured is dominant or recessive. This is because selection acts on the expression of alleles in the phenotype rather than directly on genotypes or on the alleles themselves.

An example of selection – melanism in the peppered moth

The peppered moth, Biston betularia, is found in two forms, the normal pale coloured type and a dark (melanic) form. Melanism is due to a single dominant allele.

Before the industrial revolution in Britain populations of peppered moths consisted almost entirely of the light coloured form. In terms of allele frequencies, the frequency of the recessive light colour allele was nearly 1.0.

As industrialisation killed off mosses and lichens and created dark surfaces on which moths had to hide during the day time, the selective advantage shifted towards the melanic form. Birds, the moths' main predators, could now see the light coloured moths more easily than the dark ones and ate relatively more of them.

The net result was that, in industrialised areas, the frequency of the melanic allele rose in the population until it overtook the light colour allele. The selection pressure was so great that the change in the gene pool, from almost entirely non melanic alleles to the complete reverse, occurred in probably less than fifty generations in some areas. Meanwhile, in country areas, populations remained unchanged (see fig. 49).

Balance of selection

Different selection pressures act on different aspects of phenotypes. There is, therefore, a balance of selection on any individual in a population. The result can be that an allele is maintained in a population at a higher than expected frequency as a result of selection

- homozygous, double recessive
- light coloured form

- heterozygous or homozygous
- melanic form

pre industrial revolution and country areas

industrialised areas

- light form better camouflaged
- dark form more conspicuous
- birds find dark moths more easily

- dark form now better camouflaged
- light form conspicuous
- light moths are more easily found and eaten

Result
- selection favours light form
- light form predominates

Result
- selection favours dark form
- dark form predominates

Fig. 49. Selection in the peppered moth.

for an entirely different allele with which it is temporarily or permanently linked. Alternatively, one allele may be disadvantaged by association with another.

The mixture of selection pressures on the whole phenotype of individuals leads to a 'swings and roundabouts' state of affairs which tends to maintain variety in populations.

Polymorphisms

A particularly noticeable consequence of a balance of selection pressures is the existence, in a single freely interbreeding population, of two or more distinct genetic classes. The stable coexistence of genetic types of this kind is described as polymorphism. The presence in a population of equal numbers of separate male and female individuals is a special case of a polymorphism.

Another example is the case of human sickle cell anaemia. The recessive allele causing sickle cell disease is only maintained in the population at a relatively high frequency because it imparts resistance to malaria. The homozygous recessive state produces the serious sickle cell illness, but individuals who are heterozygous for the gene in question have the advantage of being resistant to malaria. Those with two dominant alleles have perfect blood but no malarial resistance. The balance of selection between good blood and resistance to malaria maintains a balanced polymorphism for sickle cell in the population.

Mutation

You have already learned that mutation is the only way in which completely new alleles can be produced. But, can mutation significantly affect the frequencies of alleles in populations?

A typical mutation rate might be 1/100,000. This means that, for a dominant mutation, two out of every 100,000 individuals that appear each generation will have the mutant phenotype (two because each individual carries two copies of the gene). At this rate, if there were only one other allele at the gene locus in question, it would take about 70,000 generations for a mutant allele to reduce the frequency of the other allele by a half.

In other words, mutation alone has little effect on gene frequencies. It is only when selective advantage favours a mutant allele that its frequency can increase at a significant rate, as happened in the case of the allele for melanism in the peppered moth.

Migration

Populations can be reproductively isolated to differing degrees. In a Hardy-Weinberg type population there is no mating at all between members of the population and individuals from other populations. In nature it is more likely that some movement of individuals will occur between populations and some mating will take place.

You should appreciate that migration refers to migration of genes into or out of gene pools, and that what matters in population genetics is the movement of genes into new generations. Usually this is synonymous with the movement of individuals into or out of populations, but not always. If an individual or individuals migrates into a population but does not breed, the gene pool is unaffected. Or individuals may leave a population after their breeding period has finished. Again, the gene pool is unaffected. Genes may migrate into a population in gametes rather than in whole diploid individuals. Two plant populations, for example, can experience migration of genes if the pollen from one is dispersed far enough to effect pollination of plants of the other.

The effect of migration on gene frequencies, will depend on the rate of migration, and the degree of genetic difference between the migrants and the resident population.

The migration of genes in population genetics is referred to as gene flow.

Genetic drift

The Hardy-Weinberg equilibrium specifies a large (theoretically, infinitely large) population. The reason is to rule out the possibility of chance errors. The Hardy-Weinberg formulae for calculating gene frequencies are based on the assumption that, during Mendelian inheritance of alleles, genotypes are produced in their expected ratios (3 : 1 in a monohybrid cross, for example). When only small numbers of progeny are produced, however, deviations from expected ratios become quite likely.

Random deviations from expected progeny ratios in small populations can give rise to changes in gene frequencies. In some cases they can even lead to the complete elimination of alleles from populations.

Chance fluctuations in gene frequencies are known as genetic drift. Populations become vulnerable to genetic drift whenever they become small. A small part of a larger population might become isolated, or a natural disaster might reduce a population to a very small size. Whatever the cause, once a population is reduced sufficiently, genetic drift can play an important part in altering its gene pool.

Meiotic drive

Two paired alleles or, in some cases two homologous chromosomes, do not always segregate equally into gametes. The mechanism that causes preferential segregation into gametes is called meiotic drive. The result of meiotic drive is that relatively more gametes are produced that carry the favoured allele or chromosome. More progeny will then carry the favoured allele or chromosome in the next generation.

An example of meiotic drive is seen in a species of Drosophila where some male flies only produce gametes carrying X chromosomes. Normal males would be expected to produce equal numbers of X and Y carrying gametes. Male flies producing only X chromosome carrying gametes, of course, will then only have female progeny.

Tutorial

Progress questions
1. What is meant by the term 'gene pool'?

2. To what sort of population does the Hardy-Weinberg equilibrium principle apply?

3. A sample of a population of peppered moths consisted of one thousand moths, forty of which were light coloured, the rest melanic. Assuming the population fulfilled the requirements for Hardy-Weinberg equilibrium, what were the relative frequencies of the normal and the melanic alleles in the gene pool?

4. Are dominant alleles usually more frequent than recessive?

Seminar discussion
1. How would you define 'population' in the genetic sense?

2. What factors can change the frequencies of alleles in gene pools?

Practical assignments

A number of the principles applying to changes to gene pools can be investigated using models based on artificial populations made up from different coloured beads. Here is one example that illustrates the effect of selection, and mimics the case of melanism in the peppered moth.

(a) Make a population of 100 light coloured beads and 10 dark ones.

(b) Now apply selection that favours the dark beads by removing half the light beads but only one tenth of the dark ones. This represents birds eating more of the light moths because they can find them more easily.

(c) Bring the population back to approximately the starting size by adding light and dark beads without changing the proportions of each that have been established by selection. This gives the next generation.

Repeat steps (a) and (b) several times, and record the changes in the proportions of dark and light beads. Here's what you should find:

	light	dark
Starting population	100	10
after selection	50	9
new population 1	100	18
after selection	50	16
new population 2	94	30
after selection	47	27
new population 3	61	35
etc		

Try making up other games to test, for example, the effect of genetic drift, migration or a balance of selection pressures (what would happen in the above example if light coloured moths reproduced more successfully than dark ones and how could you incorporate this into the model?)

10

Genetic Technology

One minute summary – The basis of genetic technology is that DNA is the key to all genetic characters. By manipulating the DNA of cells and organisms it becomes possible to give them characteristics they previously did not have or to correct characteristics that are harmful or unwanted. The technique of changing the composition of DNA molecules is called 'recombinant DNA technology'. Recombinant DNA technology has enabled bacterial cells to produce non-bacterial gene products, to clone foreign genes, and to make gene probes. New genes have been introduced to plants and animals to make transgenic organisms. Gene probes are used for diagnosis and for genetic fingerprinting and progress is being made in the treatment of hereditary illnesses through gene therapy. In this chapter we will explore:

▶ the techniques of recombinant DNA technology
▶ some practical applications of recombinant DNA technology
▶ the uses of gene probes
▶ the problems of genetic technology that account for the gap between theory and practice

Recombinant DNA technology

Recombination, as we have found in previous chapters, involves the breaking of two pieces of DNA and their reuniting in a different combination. The DNA of one organism is, quite literally, recombined with that from a second organism. In the process genes are exchanged and new combinations of genes produced.

Scientists can now mimic the natural process of recombination in the laboratory. DNA can be extracted from one organism, cut into pieces, and the pieces inserted into the DNA of a different organism. The technology that makes this possible is referred to as recombinant DNA technology.

Theory and practice

Recombinant DNA technology is still in its infancy. It is an invaluable research tool and has contributed much to our understanding of genetics and development, but it has had relatively few successful commercial applications. Theory remains far ahead of practical exploitation.

Since all DNA is the same it should be possible, in theory, to take any gene and insert it into any host organism. Ultimately it should, again in theory, be possible to manufacture organisms according to a genetic recipe. A number of practical problems, however, slow down the realisation of this goal. It is important not to overestimate the capabilities of modern genetic technology.

The tools of recombinant DNA technology

Four basic tools are required to make and replicate recombinant DNA:

1. an enzyme which cuts DNA into pieces
2. something to carry the pieces of DNA, known as a vector
3. an enzyme to join pieces of DNA together
4. a host organism in which the recombinant DNA can replicate

Cutting DNA with enzymes

The enzymes that cut DNA into pieces are called restriction endonucleases, or restriction enzymes. The term 'restriction' comes from the fact that the natural function of these enzymes is to defend bacterial cells against attack by viruses. Bacteria which possess these enzymes can break down the viral DNA and, thereby, restrict the reproduction of the virus.

The enzymes are called endonucleases because they cut DNA within (endo = within) its linear structure rather than breaking it down bit by bit from the ends.

Restriction enzymes are used in genetic technology to cut purified DNA into manageable pieces. This is the first step in isolating particular genes.

Restriction sites

One characteristic of restriction enzymes is that they only cut DNA at certain sites where there is a particular sequence of base pairs. These sites are called restriction sites. Digestion of DNA by restriction enzymes, therefore, is not a random process. The same DNA will be broken into the same collection of fragments, each time the same restriction enzyme is used.

One particularly useful restriction enzyme cuts DNA wherever it finds the base sequence GAATTC (5′ to 3′ direction). The enzyme makes a cut in each single strand between the G and the A.

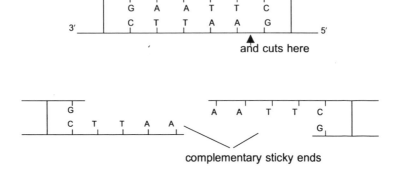

Fig. 50. Restriction sites for one particular restriction enzyme.

Notice that each strand of the DNA is cut in such a way that the cut ends are staggered. DNA fragments cut with this enzyme, therefore have a small length of single stranded DNA at each end, AATT at one end and TTAA at the other. These ends are called sticky ends. They are called sticky because they can stick to each other by base pairing. It will not have escaped your notice that the base sequences of the two types of sticky end are complementary.

Vectors

Vectors are DNA molecules that can pick up the fragments of DNA produced by restriction enzymes and carry them into a host organism. Two such vectors are plasmid DNA from bacteria and viral DNA from bacteriophage. Note that both types of vector are DNA molecules that can replicate inside a bacterial host. A fragment of DNA obtained by the action of a restriction enzyme would not be replicated inside a bacterial host unless it were combined with a vector in this way.

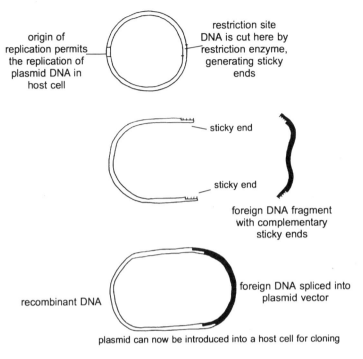

origin of replication permits the replication of plasmid DNA in host cell

restriction site DNA is cut here by restriction enzyme, generating sticky ends

sticky end

sticky end

foreign DNA fragment with complementary sticky ends

recombinant DNA

foreign DNA spliced into plasmid vector

plasmid can now be introduced into a host cell for cloning

Fig. 51. The work of a plasmid vector.

To combine DNA fragments with a vector, the vector itself is first cut by restriction enzymes. This generates sticky ends on the cut ends of the vector molecule which can then base pair with the sticky ends of the restriction fragments of the donor DNA.

Joining DNA

Sticky ends only anneal by means of hydrogen bonding between complementary bases. To attach the strands of DNA together, end to end, by chemical bonds, requires the enzyme, DNA ligase. When the cut ends of the vector molecule and the foreign DNA fragments are joined by this enzyme, the result is a piece of recombinant DNA.

▶ *Key point* – A restriction enzyme and a ligase are the only enzymes the genetic engineer needs to combine two pieces of DNA from different sources. You can think of them as the scissors and glue of genetic engineering.

A host organism

The purpose of a host organism, initially, is to replicate the DNA that has been recombined into a vector molecule. The most widely used host in genetic engineering is the bacterium E. coli, though others such as yeast and mammalian cells are sometimes employed. One advantage of using E.coli is that it will take up vector DNA and any foreign DNA the vector is carrying by the same process as happens in transformation.

Once the vector, along with the foreign DNA it carries, is incorporated into the host cell, it replicates as the host cells multiply. As many copies of the foreign DNA are made as there are host cells produced. In the case of E. coli as host this could be many millions in a short time. Because the genes in the foreign DNA are being replicated inside the host, the process is referred to as gene cloning and the vector is called a cloning vector.

As well as replicating a gene, a host organism can, in a few instances, be made to synthesise the gene product.

How genes are isolated

Using the tools described above it is possible to get pieces of foreign DNA to replicate inside a bacterial cell. It is a huge step from there to being able to clone a particular selected gene.

A particular gene can be selected for cloning in three ways. It can be made to order, made by reverse transcription, or isolated by 'shotgunning'.

The gene is made to order

If the amino acid order of a gene product is known, a sequence of nucleotides can be assembled in the laboratory that will code for that protein. Such an artificial gene can then be attached to a suitable vector and introduced to a host.

Most proteins, however, are too large to make the synthesis of an appropriate gene a realistic proposition.

The gene is made by reverse transcription

In cells where a particular gene is extremely active there is a lot of messenger RNA produced by that gene. For example, in cells of the pancreas that secrete insulin there will be a lot of insulin mRNA. It is easier to isolate this mRNA than to find the gene that made it.

Using the mRNA it is possible to reverse the transcription process and make DNA from an RNA template. The DNA will be complementary to the mRNA and have the same base sequence as the gene in question. Reverse transcription is made possible by the enzyme reverse transcriptase which operates naturally in a group of viruses known as retroviruses.

▶ *Key point* – Making a gene by reverse transcription is a method which has been used for the production of human insulin by ·bacterial hosts.

The gene is isolated by 'shotgunning'

The principle of the 'shotgunning' technique is to clone all the restriction fragments produced by digestion of a full complement of an organism's DNA and then to select those hosts carrying the required gene from all the others. If, for example, a complete human genome is digested with a restriction enzyme, several hundred thousand fragments of DNA are produced. One of these should carry the gene that is wanted. If the host carrying this fragment can be found the gene can be cloned.

To select bacteria that carry the required gene from a population in which the vast majority do not, may seem an almost impossible task. Two powerful selection procedures make it possible.

1. *Using a designer vector.*
A plasmid vector can be made which incorporates genes for antibiotic resistance. If antibiotics are incorporated into the media used to grow the bacteria it is possible to select, not only those bacteria carrying the plasmid, but also those carrying recombinant DNA. Bacteria carrying the required gene will be amongst this group. The rest can be discarded.

2. *Using a gene probe*
A gene probe is a piece of radioactively labelled single stranded DNA with a specific base sequence. It will anneal to any other piece of single stranded DNA that has a complementary base sequence, by means of hydrogen bonds. A gene probe, therefore, can be used to identify pieces of DNA with a particular base sequence.

In trying to find the gene he is looking for, the genetic engineer has to isolate one particular bacterial colony from several thousand colonies of bacteria carrying different types of recombinant DNA.

(a) The colonies are transferred to a special membrane and grown there until there is enough DNA in each colony for the next step in the process of gene identification.

(b) The DNA from each of these colonies is released from the bacteria, denatured and fixed to the membrane. The result is a membrane dotted with areas of single stranded DNA. Each area of DNA came from a known bacterial colony. Somewhere amongst all the different areas of DNA will be the gene to be isolated.

(c) A gene probe which has a base sequence complementary to the gene in question is added to the membrane and sticks to those places on the membrane where the gene is present. The position of the gene probe shows up on an autoradiograph.

(d) The source of the DNA to which the gene probe sticks can then be traced to the bacterial colony from which it came. The bacteria from that colony are grown on and the selected gene cloned.

Gene selection by this method depends entirely on the ability to produce a suitable gene probe. This, in turn, depends on some knowledge of the base sequence of the gene.

Gene expression in recombinant DNA technology
Even if a gene can be cloned it does not follow that the anticipated gene product will be obtained. There are three main reasons for this.

1. *Introns and exons*
Eukaryotic genes contain two types of DNA, known as introns and exons. Exons contain the code that ultimately translates into the protein of the gene product. Introns contain what is sometimes referred to as 'junk' DNA because it doesn't code for anything. A single gene may contain several exons, interspersed with introns. In the eukaryotic cell both exons and introns are transcribed into mRNA but the 'junk' RNA is later removed by the process referred to as post transcriptional modification. Only the mRNA from the exons is then translated into protein (see fig. 52).

Bacteria do not have the ability to carry out post transcriptional modification. If, therefore, any gene with both introns and exons were cloned in a bacterial host, no gene product would be produced.

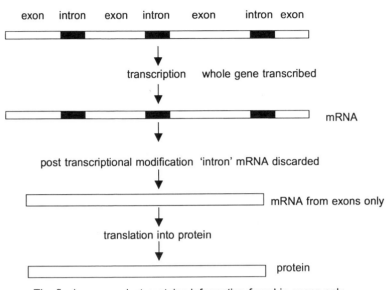

The final gene product contains information found in exons only.

Fig. 52. The operation of a 'split gene'.

2. *Post translational modification*

The initial protein made by a gene is sometimes modified to make the final functional gene product. The human insulin gene, for example, is first translated into a single chain polypeptide which is then modified to form the shorter double chain insulin molecule. Bacteria do not have the enzymes to do this post translational modification.

3. *Switching the gene on*

In bacteria genes are controlled by the operon system (see chapter 3) and switched on by a promoter. A foreign gene recombined into a bacterial plasmid will not have such a regulatory system and will, therefore, not be switched on.

Transgenic plants and animals

▶ *Key point* – Transgenic organisms are ones into which functioning foreign genes have been inserted. Such organisms are also referred to as genetically modified plants and animals.

We have seen how genes can be isolated, cloned and made to function in bacteria. The human insulin gene, for example, can be cloned in cells of E. coli and the hormone itself produced by the bacterium. The same principles are employed in making transgenic plants and animals but there are additional problems to overcome.

Problems for making transgenic plants and animals

1. Eukaryotic cells do not always take up DNA as readily as bacteria, nor do they have plasmids.

2. The DNA in cells is combined with protein in chromosomes.

3. Plants and animals are multicellular.

4. For the new genetic character of a transgenic organism to be inherited it must be incorporated into the germ line as well as into somatic cells.

▶ *Key point* – Despite these additional problems it has proved possible to make transgenic multicellular organisms in which the new genes are inherited according to the same Mendelian principles that apply to the rest of their genome.

Making a transgenic plant

Plants offer the advantage to the genetic engineer that many of them can be reproduced from single cells by tissue culture methods. If, therefore, a single cell has a gene or genes inserted, a whole transgenic plant can often be generated.

Inserting DNA into plant cells

Once the chosen gene has been isolated and cloned there are a number of ways it can be introduced into a plant cell. Inside the cell, the gene is usually integrated into the host genome without difficulty.

Foreign DNA can be introduced into plant cells by the following methods. Each method has been found to be able to bring about the incorporation of the new DNA into a host genome.

1. Direct application of pure DNA – an analagous method to bacterial transformation.

2. Firing minute DNA coated particles of tungsten or gold into the cell from a particle gun – a method known as biolistic transformation.

3. Direct microinjection of DNA into cells.

4. By infecting plants with viruses carrying the foreign DNA.

5. By using the tumour forming bacterium Agrobacterium tumefaciens as a carrier.

This last method is one of the most widely used for introducing foreign genes into plants. The bacterium Agrobacterium tumefaciens is a soil organism which can infect plants and cause abnormal growths. Its ability to do this depends on the presence of a tumour inducing plasmid, part of which becomes integrated into the host genome when the bacterium infects a plant. If foreign genes are inserted into the plasmid by genetic engineering they are also recombined into the DNA of the invaded plant cells.

Commercial applications of transgenic plants

Genetic modification has been successfully employed on a commercial basis to produce crop plants which have added characters such as: herbicide resistance, insect resistance, virus resistance, frost hardiness and controlled ripening.

Transgenic mammals

The main problem in making transgenic mammals is that they have to develop from a fertilised egg. They cannot be reproduced in any other way. If new genes are to be introduced, therefore, it must be at an early stage in development.

Some success in producing transgenic mammals has come using methods of introducing foreign DNA similar to those used in the production of genetically modified plants. Transgenic mice, for example, can be made by injecting foreign DNA directly into fertilised eggs and replacing the eggs into a foster mother. A small proportion of the injected eggs develop into mice which show the phenotype associated with the introduced gene and can pass it on to their offspring.

Genetic fingerprinting

Genetic fingerprinting is, as its name suggests, a method for identifying individuals from their DNA. It can also be used for establishing family relationships. The success of the method relies on two facts:

(i) The DNA is cut into restriction fragments by a restriction enzyme: the number and size of these fragments are unique to the individual.

enzymes cut the DNA at restriction sites

(ii) Restriction fragments are sorted according to size by gel electrophoresis. The fragments distribute themselves down the gel, smaller at the bottom and larger at the top.

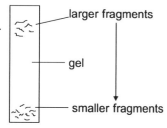

larger fragments

gel

smaller fragments

(iii) The DNA fragments in the gel are denatured (made single stranded) and transferred to a membrane, without altering their relative positions.

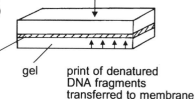

membrane   gel   print of denatured DNA fragments transferred to membrane

(iv) A radioactive gene probe is added which is designed to hybridise with repeating sequences in the denatured DNA. Wherever there is a fragment of DNA containing such a sequence on the membrane the probe will identify its position. The relative positions of the various sized fragments is shown up on an autoradiograph as a series of bands. The pattern of the bands is unique for every individual.

autoradiograph showing DNA fragments as bands

Fig. 53. Genetic fingerprinting.

1. When human DNA is digested by restriction enzymes the number and size of fragments are unique to an individual. You will remember that restriction enzymes cut DNA at specific sites where they recognise certain base sequences.

2. The human genome contains several non-coding regions of DNA (introns or 'junk' DNA) consisting of short sequences of bases repeated many times. These sequences are found throughout the genome and a gene probe can be made to identify them.

The procedure for producing a genetic fingerprint is shown in figure 53 and an example of the result of a genetic fingerprint test is shown in fig. 54.

Fig. 54. Part of DNA fingerprints from two related individuals and one unrelated person showing similarities only between the two related prints.

Medical applications of recombinant DNA technology

Diagnosis

1. *Infectious diseases*
Gene probes can be used for the rapid diagnosis of bacterial and viral illnesses. All that is needed is a radioactive probe that is complementary to part of the DNA of the infective organism. If the probe hybridises with any denatured DNA taken from a tissue sample, the DNA identified must be from the disease organism and the tissue is shown to be infected.

2. *Hereditary diseases*
Restriction enzymes and suitable gene probes are also used for the early detection of hereditary diseases such as sickle cell anaemia, Duchenne muscular dystrophy, cystic fibrosis and Huntingdon's chorea and for the identification of carriers of recessive genes for such diseases.

The identification of individuals heterozygous for sickle cell anaemia, for example, depends on the fact that the restriction fragments produced by digesting DNA containing the normal gene are of different lengths to those obtained from DNA that has the recessive, sickle cell allele. The size difference can be shown up by electrophoresis of the DNA fragments and the site of the sickle cell gene detected with a gene probe. If the gene probe hybridises with a small fragment of DNA it indicates the presence of the normal allele. If it hybridises with a larger fragment it means a recessive, sickle cell allele.

Gene therapy

The treatment of hereditary diseases by replacing bad genes with good ones, known as gene therapy, is a realistic goal in the medical application of genetic technology. As yet, work in this direction is still at the research stage. Perhaps the biggest problem for genetic modification of this sort is the fact that we are multicellular organisms. A genetic disease might be treated by introducing normally functioning genes to some cells of the body but it is not possible to change the whole genotype of an individual unless it is done at a very early developmental stage.

There is much ethical resistance to the manipulation of human eggs and early embryos. For this reason the progress of gene therapy is primarily in the direction of somatic gene therapy; this treats only some cells of the body, and does not affect the germ line. The results of any genetic change to somatic cells would not be passed on to the next generation.

Tutorial

Progress questions
1. What role do the following play in recombinant DNA technology

   (a) restriction enzymes
   (b) DNA ligase
   (c) cloning vectors

2. What is meant by 'shotgunning'?

3. If a gene can be cloned inside a bacterial cell, why does it not automatically produce its normal protein product?

4. What is a gene probe and how is it used?

Seminar discussion
What are the ethical problems associated with:

(a) transgenic organisms
(b) pre-natal diagnosis of hereditary diseases
(d) genetic counselling
(e) gene therapy

Practical assignments
1. One thing that has been achieved is the bacterial synthesis of human insulin. Try to find out the full story of this successful piece of genetic engineering.

2. Another success story is the production of transgenic plants and their incorporation into genetically modified food products. Why has there been considerable resistance to this development and why have such plants and their products been banned in some places? See if you can find out.

Study tips

1. In discussing the implications of genetic technology, and arguing for or against any of its applications, it is vital to define the nature of any argument. Is it ethical, practical, economic, hypothetical or simply uninformed and emotional? Remember, also, that there is a distinction between a technique and its application.

2. Genetic technology is a rapidly moving field in which possibilities far outweigh actual results. It is also heavily dependent on advanced and hard to understand techniques. You would be well advised to concentrate only on the basic theory and on those practical applications that have already proved successful.

Appendix

Answers to progress questions

Answers to progress questions are only given here where they are of a numerical or otherwise clear cut nature and cannot be unambiguously obtained by reading the text.

Chapter 2
Question 2
(a) The base sequence of the other strand is: TAACTGCCTTGAACG.
(b) The direction is opposite to that of the other strand.
(c) The two strands are complementary.

Chapter 3
Question 3
Sequence of amino acids =

met leu thr val lys trp (term)

Chapter 4
Question 2
Phenotype =
    Unable to metabolise lactose.
    Requiring external supply of leucine for growth (leucine auxotroph).
    Sensitive to streptomycin.

Question 3
(a) Because there is no leucine in the medium.

(b) Because it is unable to use the lactose as an energy or carbon source and there is no alternative.

(c) Because it cannot grow in the presence of streptomycin.

Chapter 7
Question 2
Yes they are linked – because there is no recognisable Mendelian ratio amongst the progeny and parental types predominate.

Alleles are linked: tall to round and dwarf to pear (dominant to dominant and

189

recessive to recessive).

map distance $= \dfrac{23 + 17}{200} \times 100 = 20$ map units

Question 3
If only one cross over occurs the maximum value for recombination is 50%, giving a map distance of 50 units. This is because recombination is semi-conservative, involving only two of four chromatids at each cross over. Actual map distances between genes can be any magnitude. This is because they are calculated by adding many smaller distances together. Over longer distances the chance of there being more than one cross over becomes quite large.

Chapter 8
Question 4
(a) the DNA codes for: met trp val.

(b) i. ile would be substituted for met.

   ii. ATC is a termination code and the transcription would stop at this point.

   iii. There would be no change: both CAC and CAT are DNA codes for valine.

(c) If the T were deleted the message would read – cys gly – but would it start in the absence of an initiator?

Chapter 9
Question 3
If the frequency of the normal allele is q and that of the melanic is p

$$\text{then } q^2 = \frac{40}{1,000} = \frac{1}{25}$$

$$q = \frac{1}{5}$$

$$p + q = 1$$

$$p = \frac{4}{5}$$

Glossary

acridine dyes Chemical mutagens causing frame shift mutations.

adenine One of the bases of DNA and RNA: adenine pairs with thymine in DNA and uracil in RNA.

adaptive enzyme An enzyme which is only produced in the presence of its substrate.

agar A jelly like substance used for making solid media for growing bacteria.

alkylating agent Chemical mutagen that works by converting one DNA base to another.

allele One particular form of a gene.

allele frequency The relative frequency of an allele amongst genotypes in a population; also called gene frequency.

amino acid the basic building block of proteins, occurring commonly in nature in twenty varieties.

amino group That part of an amino acid which contains nitrogen and combines with the carboxyl group of a second amino acid in the formation of protein.

aneuploidy having more or fewer chromosomes than the normal number but not an exact multiple of sets (see **polyploidy**).

annealing Single strands of DNA joining together by means of hydrogen bonds between bases to form double stranded DNA.

antibiotic a substance with antibacterial action.

anticodon a triplet of bases found on transfer RNA which pairs specifically with the codon on the messenger RNA, bringing the correct amino acid to the site of protein synthesis.

ascospore a spore produced by an ascomycete fungus, usually as one of a group of eight, arranged linearly in a spore case called an ascus.

ascus the spore case of those fungi belonging to the group, ascomycetes.

asexual reproduction reproduction that is not preceded by meiosis and sexual fusion and does not involve genetic mixing.

autoradiograph a picture formed by the exposure of a photographic plate to a radioactive source.

autosome a chromosome which is not a sex chromosomes.

auxotroph a microorganism that cannot grow without the addition of a particular nutrient to the medium on which it is living.

back cross a cross between a parent and one of its progeny.

bacterial lawn a continuous even growth of bacteria on solid medium used for the detection of plaques formed by bacteriophages.

bacteriophage a virus parasitic on bacteria.

bacterium a prokaryotic organism of the kingdom Monera.

base one of the molecular components of nucleic acids. There are four types found in DNA: adenine, guanine, cytosine and thymine and four types commonly found in RNA: adenine, guanine, cytosine and uracil. Much of the function of nucleic acids depends on the pairing between bases.

191

base analogue chemicals that have similar structure to the bases of nucleic acids and can substitute for those bases, thereby causing mutations.

base equivalence rule the rule that states that in double stranded DNA the molecular amounts of the bases adenine and thymine should be equal and those of guanine and cytosine should be equal; the reason lies in the specificity of base pairing, A with T and C with G.

base pair two opposite bases in double stranded nucleic acid joined by hydrogen bonds.

base sequence the linear sequence of bases in a single strand of nucleic acid, usually given as a series of letters, reading in one particular direction.

binary fission reproduction of unicellular organism by cell division.

biolistic transformation the transfer of foreign DNA into a cell of a higher organism by firing minute DNA coated projectiles into the cell.

biosynthesis the manufacture in a living system of a chemical product by enzymatically controlled reactions.

biosynthetic pathway the sequence of enzymatically controlled steps leading to an end product of biosynthesis.

bivalent four paired chromatids visible at first metaphase of meiosis.

blender a piece of laboratory apparatus equivalent to a kitchen liquidiser used to macerate tissue and, famously, to separate virus particles from the bacteria they had infected in the Hershey-Chase experiment.

breakage and reunion the mechanism of crossing over and recombination: both strands of DNA break and reunite in a different combination.

burst size the number of viral particles released on lysis of a single bacterium infected by a bacteriophage.

carboxyl group the carbon, hydrogen and oxygen containing part of an amino acid which combines with the amino group of a second amino acid in the formation of protein.

catalyst a material which speeds up a chemical reaction without itself being altered; enzymes are biological catalysts.

cell a unit of biological organisation, bounded by a membrane and containing nucleic acid. Cells may be parts of multicellular organisms or whole, single celled organisms. Eukaryotic cells have a nucleus. Prokaryotic cells do not.

cell division the forming of two cells, or four in the case of reduction division, from one. During cell division the nucleic acid of the cell is replicated and divided equally into the daughter cells.

cellulose The material making up the cell wall of many plant cells.

central dogma the idea that genetic information passes from DNA to RNA to protein.

centromere a constriction in a chromosome that plays an important role in the movement of chromosomes and chromatids during cell division.

chiasma the visible sign of crossing over in homologously paired chromosomes.

chi-square test a statistical test of the fit of an observed ratio to a predicted ratio – much used in testing whether progeny ratios in genetic crosses fit known Mendelian ratios.

chloroplast the photosynthetic apparatus of plant cells: a membrane bound organelle containing some DNA.

chromatid a newly formed chromosome, still held to its sister chromatid at the centromere.

chromosome the structure of eukaryotic cells which carries DNA. Chromosomes are found in the nucleus and are visible during cell division. Genes are arranged linearly along the chromosomes.

chromosome mapping establishing the order of genes along a chromosome and their relative distances apart by means of genetic experiments.

classical genetics that branch of genetics dealing with experimental crosses involving higher organisms.

clone individuals produced by asexual reproduction that are, therefore, genetically identical.

cloning vector a piece of DNA to which can be attached a foreign gene in order that the gene can be introduced to a host cell and replicated (cloned). Bacterial plasmids are often used as cloning vectors.

codominance the state when neither of two alleles is dominant over the other: they are equally or co-dominant. Individuals heterozygous for the codominant alleles show *both* characteristics.

codon a triplet of bases on messenger RNA which provides the code for an amino acid or a punctuation point in the synthesis of protein; the codon pairs specifically with the anticodon of one species of transfer RNA.

colinearity of gene and protein the finding that a mutation in one part of a gene produces an amino acid change in an equivalent part of the protein that the gene is responsible for making; the linear sequence of a number of mutations in the gene mirrors the linear sequence of amino acid changes in the protein, thus confirming the concept of a genetic code.

colony a visible mass of bacterial cells growing on solid growth medium; all the cells of the colony are the product of cell division from a single progenitor and are, therefore, genetically identical.

complementary of nucleic acid base sequences: such that they can pair according to the rules of specific base pairing: A with T or U and C with G.

complex virus a virus with a relatively complex structure, for example, bacteriophages.

conjugation the process by which bacteria join together and exchange genetic material.

conjugation bridge a tube joining two conjugating bacteria, across which DNA can pass from one cell to the other.

conjugation tube an alternative name for conjugation bridge.

cotransduction simultaneous transfer of two or more genes from one bacterium to another by means of a transducing virus.

cross fertilisation fertilisation of one organism by another distinct individual.

crossing over the exchange of genes by breakage and reunion of homologous chromosomes.

cytoplasmic inheritance inheritance of non chromosomal DNA, usually solely through the maternal line, via organelles in the cytoplasm.

cytosine one of the bases of DNA and RNA: cytosine pairs with guanine.

deletion chromosomal mutation involving the loss of a portion of the chromosome.

denaturing rendering double stranded DNA into its component single strands

deoxyribose the sugar component of DNA.

deoxyribonucleic acid DNA – the material of inheritance.

differentiation the process by which cells and tissues change during development towards some specialisation.

dihybrid cross a cross in which the parents differ by two distinct characteristics.

diploid having two complete sets of chromosomes, twice the haploid number; or, of part of the genome, having two copies of a homologous region.

DNA deoxyribonucleic acid.

DNA ligase see ligase.

DNA polymerase enzyme catalysing the synthesis and breakdown of DNA.

DNA repair the conversion of erroneous DNA into its correct form.

DNA replication the formation of two identical copies of DNA from one source molecule by a semi conservative process.

DNA sequencing establishing the order of nucleotide bases in a piece of DNA.

dominant of an allele: exerting its influence in the heterozygous state over and above that of the other allele (the recessive).

dosage effect a change in the resulting phenotype expressed by an allele depending on how many copies of the allele are present in the genotype.

double crossover two cross overs involving the same chromatids, effectively cancelling each other out.

double helix the postulated three dimensional shape of the double stranded DNA molecule.

Down's syndrome a human condition caused by the presence of an extra chromosome.

duplication Chromosomal mutation in which part of the chromosome is repeated.

E. coli the bacterium *Escherichia coli*, much used in genetic research.

egg cell the female gamete: a haploid cell which joins with the male gamete at fertilisation.

electrophoresis a means of separating molecules and parts of molecules on a gel by means of an electric poential difference; a very useful analytical technique.

endonuclease see **restriction endonuclease.**

enzyme a biological catalyst; all biochemical reactions in cells are believed to be controlled by enzymes and, generally speaking, a different enzyme controls each separate reaction.

enzyme repression the prevention of synthesis of an enzyme due to the presence of a product of the reaction that the enzyme controls.

epistasis interaction betwen non allelic genes such that the effects of one pair are masked or modified by the other pair.

eukaryote an organism having a cellular structure including a membrane bound nucleus, cell organelles and DNA in chromosomes; the alternative to **prokaryote.**

evolution the process of change in living organisms leading to the production of new species.

exons regions of the DNA of eukaryotic organisms which code for parts of functional proteins – as opposed to **introns.**

F1 (first filial generation) the progeny of a cross between two parents both homozygous for particular character differences; the first progeny that Mendel obtained in his crosses.

F2 (second filial generation) the progeny of a cross between two parents both heterozygous at a particular gene locus; the second progeny that Mendel obtained from his crosses.

F factor see **fertility factor.**

F plasmid a piece of DNA, additional to the bacterial genome, either free in the bacterial cell or incorporated in the bacterial chromosome, which gives a bacterium the ability to act as a donor of genetic material by conjugation.

female the sex that makes large immobile gametes, usually known as eggs or egg cells.

fertilisation the process by which male and female gametes join to form a zygote.

fertility factor an alternative name for the **F plasmid** of bacteria.

frame shift mutation a mutation caused by the loss or addition of nucleotide bases in DNA; the result of such a mutation is a shift in the reading of the genetic code such that its whole meaning is lost.

fungal cellulose the cellulose-like material making up the cell walls of fungi.

galactose a simple sugar, one of the two monosaccharides, along with glucose, making up the disaccharide lactose.

galactosidase the enzyme required to break down lactose to its component monosaccharides.

gamete a usually haploid cell which fuses with a second gamete during fertilisation to restore the diploid phase of the life cycle.

gene in classical genetics: a unit of inheritance that behaves according to Mendel's principles; in molecular genetics: a length of DNA responsible for the production of a single protein.

gene cloning the production of a large number of identical copies of a gene or other piece of DNA introduced into a cloning host (usually a bacterium) by the techniques of genetic engineering.

gene flow the movement of genes within or between populations.

gene frequency see **allele frequency.**

gene interaction the situation where two or more pairs of alleles affect the same phenotypic character.

gene locus the position on a chromosome at which a gene is found.

gene pool the sum total of all the genes in a population.

gene probe a single stranded piece of DNA which will hybridise with a particular DNA base sequence and thereby identify it; gene probes are usually radioactive for ease of recognition.

gene product the particular protein, usually an enzyme, encoded in a gene and made by it.

gene repression the inhibition of the action of a gene by a repressor substance, in many cases the gene product.

gene therapy the treatment of inherited diseases by genetic engineering approaches.

genetic code the code that lies in the sequence of bases in DNA and translates into proteins.

genetic drift changes in the frequency of genes in populations due to chance factors.

genetic engineering manipulation of the DNA of organisms for the purposes of, for example, changing inheritance or the *in vitro* manufacture of gene products.

genetic equilibrium the state of a population when gene frequencies do not change from generation to generation.

genetic fingerprint a pattern of DNA fragments, unique to the individual, identified by means of a gene probe.

genetic map see **mapping**.

genetic marker any point of variation that allows a distinction to be made between individuals on genetic grounds.

genetic modification the alteration of the genotype of a cell or organism by genetic engineering.

genetic technology the whole field of DNA manipulation.

genetic transformation altering the genotype of a cell or organism by application of pure DNA; genetic transformation was first found to be effective with bacteria and most commonly applies to bacterial cells.

genetic variability the potential to vary genetically in future generations.

genetic variation variation between individuals or populations which has a genetic basis and is, therefore, hereditary.

germ cell a gamete.

germ line the ancestral line of cells from which gametes are derived, ultimately going back to the zygote from which an organism developed.

GM abbreviation for **genetically modified.**

guanine one of the bases of DNA and RNA: guanine pairs with cytosine.

Hfr high frequency: referring to the high frequency with which recombinants appear in crosses between bacteria, one of which is an Hfr strain.

haploid having only one set of chromosomes.

Hardy-Weinberg equation an equation enabling gene frequencies to be calculated from observed phenotype frequencies.

Hardy-Weinberg equilibrium the condition of a population in genetic equilibrium when the Hardy Weinberg equation applies.

heavy nitrogen the isotope of nitrogen with one additonal neutron in each atom (^{15}N) used as a marker for DNA in the Meselson-Stahl experiment.

hermaphrodite having both male and female sex organs.

heterogametic producing two types of gamete, one with an X chromosome, the other with a Y chromosome.

heterogametic sex the sex, male or female, that produces two types of gamete

heterosis hybrid vigour.

heterozygous having two different alleles at one gene locus.

heterozygosity the state of being heterozygous.

histones the proteins associated with DNA in chromosomes.

homogametic producing only one type of gamete with respect to the sex chromosomes.

homogametic sex the sex, male or female, that produces only one type of gamete

homologous pairing the pairing of chromosomes at meiosis, homologous region to homologous region.

homology referring to chromosomes or DNA: having similar and related but not necessarily identical structure: having the same genes in the same order but not necessarily the same alleles.

homozygous having two identical alleles at one gene locus.

hormones substances that have a profound effect on development and biological control and can be implemental in the switching on of genes.

host range mutant a mutant of bacteriophage which allows the phage to infect previously immune strains of bacteria.

hybrid an individual resulting from a cross between two genetically different parents: any individual that is heterozygous.

hybrid DNA DNA that results form the annealing of single stranded DNA from different sources.

hybrid vigour growth or other vigour that characterises heterozygous individuals: also called heterosis.

hybridisation the crossing of two genetically different individuals to obtain hybrid progeny.

hybridisation (of DNA) annealing of single strands of DNA from different sources to form a piece of double stranded hybrid DNA.

hydrogen bonds forces of attraction holding molecules together; base pairing in nucleic acids is by means of hydrogen bonds.

hydroxy-methyl-cytosine a base found in place of cytosine in bacteriophage DNA.

inbreeding breeding between individuals that are themselves related.

incomplete dominance the situation where neither of two alleles is completely dominant over the other in its phenotypic effect so that a heterozygous individual has a different phenotype to either homozygote.

incomplete sex linkage of a gene that is located on both the X and the Y sex chromosomes; an incompletely sex linked gene can recombine by crossing over.

independent assortment Mendel's principle that, in a dihybrid cross, the segregation of one pair of alleles into gametes is unaffected by the segregation of the other pair. The term can latterly be applied to the independent movement of chromosome pairs at meiosis.

indicator medium a bacteriological growth medium that, by virtue of its composition, can identify certain bacterial types.

inducer a substance which switches on a gene or genes, thus inducing the synthesis of gene products.

inducible enzyme an enzyme which is produced by a gene that is switched on by an inducer substance, often the substrate for the enzyme.

insulin the hormone responsible for the control of blood sugar levels in mammals

interphase the phase during which cells are not dividing and chromosomes are not visible.

intron the non-coding parts of the genes of higher organisms inserted between the **exons.**

inversion chromosomal mutation in which part of the chromosome is turned upside down.

isotope a version of an element that has a slightly larger or smaller atom; isotopes are often radioactive and prove valuable research tools.

isotopic labelling using isotopes to distinguish between otherwise identical chemical compounds as an aid to research, diagnosis or identification.

junk DNA DNA that does not code for any final gene product and, therefore, appears to have no purpose. **Introns** are sometimes called junk DNA.

lactose a disaccharide found in milk.

latent period referring to the life cycle of a bacteriophage: that period following infection during which new phages are being synthesised in the bacterial cell but are not yet fully formed.

lethal allele an allele which can only exist in the heterozygous state since zygotes homozygous for the allele are not viable.

ligase an enzyme that joins pieces of DNA end to end.

linkage the physical association of different genes on the same piece of DNA or the same chromosome.

linkage group all the genes that are linked together on one piece of DNA or one chromosome; an organism has the same number of linkage groups as its haploid chromosome number.

linked genes genes that are linked on the same piece of DNA or on the same chromosome.

locus see **gene locus.**

lysis the breaking down of a cell, as when new phage particles are released from a bacterial cell following viral infection.

lysogeny the incorporation of bacteriophage DNA into a bacterial chromosome so that the virus is non virulent and its DNA is replicated along with that of the bacterium

male the sex that produces the smaller, mobile type of gamete.

map unit a statistical and non dimensional measure of the distance apart of genes on a genetic map.

mapping plotting the position and relative distances apart of genes on chromosomes or lengths of DNA; making a genetic map.

marker see **genetic marker**.

maternal inheritance inheritance solely through the female line i.e. from mother to progeny but never from father to progeny.

meiosis cell division in which the number of chromosomes per cell is halved and genetic mixing takes place; also called **reduction division.**

meiotic drive a distortion of meiosis through which some genetic elements appear more frequently in the meiotic products than would be expected.

melanism darkness of phenotype caused by excess of the pigment **melanin.**

melting the conversion of double stranded DNA to single strands by heating; heat breaks down the hydrogen bonding holding the two complementary strands of DNA together.

Mendel Gregor Mendel was the founder of classical genetics with his seminal experiments on inheritance in the garden pea.

Mendelian inheritance inheritance that follows the principles outlined by Mendel.

Mendel's principles the principles proposed by Mendel to explain the results of his crosses, namely: (a) pairs of alleles, (b) dominance and recessiveness, (c) segregation, (d) independent assortment.

merozygote a partial zygote that is diploid for only part of its genome, as formed, for example, following conjugation in bacteria.

messenger RNA the type of RNA that is transcribed from DNA as the first step in the production of protein.

metaphase that stage of mitosis or meiosis during which the chromosomes are assembled across the middle of the cell; in first meiotic metaphase the chromosomes are paired at this stage.

micron one thousandth of a millimetre; one millionth of a metre.

migration referring to genes: the movement of genes from one population to another

minimal medium a bacteriological medium which contains the bare essentials for growth and no additional nutrients.

mitochondrion the organelle of eukaryotic cells responsible for aerobic respiration

mitochondrial DNA the small amount of DNA found in mitochondria and controlling some aspects of mitochondrial inheritance.

mitosis replicative cell division.

molecular genetics that branch of genetics concerned with the molecular basis of genetic mechanisms.

monohybrid cross a cross in which the parents differ with respect to a single character only.

monozygotic twin a twin originating from the division of a single zygote, therefore a clone of the other twin.

mRNA abbreviation of **messenger RNA.**

mutagen any influence which causes mutations.

mutation (a) a spontaneous or induced change in the genetic material, (b) a change in phenotype arising as a result of mutation, (c) the process of mutation.

multicellular of an organism: made up of more than one cell.

multiple alleles three or more alleles of the same gene.

multiple genes more than one gene affecting the same characteristic of phenotype.

non-disjunction a mistake in cell division resulting in an uneven distribution of numbers of chromosomes to each daughter cell.

nonsense codon a codon which does not code for any amino acid; it is now known that such codons are the full stops in the genetic code; also called **terminator codons**.

nuclear membrane the double membrane surrounding the nucleus of eukaryotic cells, separating the nucleus from the cytoplasm.

nucleic acid the material of inheritance: **DNA** and **RNA**.

nucleoside a component of nucleic acid consisting of a base linked to either ribose or deoxyribose.

nucleotide a nucleoside to which is attached a phosphate group.

nucleus the part of a eukaryotic cell containing the chromosomes.

one gene – one enzyme the theory originated by Beadle and Tatum that states that the job of each gene is to produce one specific enzyme.

one-step growth curve the form of curve obtained by plotting number of infective particles against time for an infection mixture of bacteriophage and bacterial host; following a latent period during which there is no change in numbers comes a sudden large increase.

operator part of the DNA which, by interacting with a repressor substance controls the activity of a gene or system of genes; typically the operator switches on the operon system of bacteria.

operon a series of linked genes which act as a unit in the bacterial genome and are controlled by a single operator region

organelle one of several types of eukaryotic cell components, each with a particular function, for example chloroplast, mitochondria.

osmotic shock a sudden change in osmotic environment which can cause cellular disruption.

outbreeding breeding between individuals that are not related.

^{32}P a radioactive isotope of phosphorous.

pairing see **homologous pairing**.

parental type the same genotype or phenotype as the parent.

pathway analysis deciphering the sequence of reactions in a biochemical pathway by means of nutritional and genetic experiments.

peptide see **polypeptide**.

peptide bond the bond formed between the amino group of one amino acid and the carboxyl group of another during the linking of amino acids into peptides and proteins.

phage short for **bacteriophage**.

phenotype the character of an organism as it is expressed; different genotypes may have the same phenotype.

phosphate group part of a nucleotide consisting of a phosphorous atom and four oxygen atoms.

plaque the visible sign of the presence of a centre of phage infection on a bacterial lawn: the plaque is a clear space in the lawn within which all bacteria have been destroyed.

plasmid a small piece of DNA in a bacterial cell, separate from the main bacterial chromosome but potentially able to combine with it in many cases; when free of the bacterial chromosome the plasmid is circular in form.

point mutation a mutation of DNA at a confined spot, generally a single nucleotide base.

pollen grain the structure that carries the male gametes in all seed plants.

polygenes more than one gene having, collectively, an additive effect on the same character of phenotype.

polymorphism the existence of two or more distinct phenotypes in a population

polypeptide a sequence amino acids that cannot yet be referred to as a protein; may also simply be called a peptide.

polyploid having more than two complete sets of chromosomes.

population a group of individuals that tend to breed amongst themselves

population genetics the study of genes in populations.

position effect the influence of the position of a gene in relation to other genes in determining its effect on the phenotype.

post transcriptional modification a change made to mRNA prior to its translation into protein; this is often a case of the removal of that part of the RNA transcribed by introns, leaving only exon-transcribed mRNA.

post translational modification a change made to the polypeptide product of mRNA translation prior to the appearance of the final protein product.

primary structure of a protein: the amino acid sequence.

prokaryote an organism that is not eukaryotic, having no proper nucleus, no chromosomes and no cell organelles: bacteria.

promoter a region of a gene to which RNA polymerase binds as the first step in transcription.

prophage the name given to the DNA of a bacteriophage when it is combined into the DNA of a bacterium; see **lysogeny**.

protein a compound formed by the linking of many amino acids of different types and the coiling and folding of the resulting molecule to achieve a characteristic three dimensional structure which gives the molecule its essential properties.

protista the kingdom of single celled organisms that live mostly in water and used to be grouped with either the protozoa or the algae.

prototroph a microorganism that can cater for its own nutrition without the addition of any particlar growth requirement and that can, therefore, live on minimal medium; the opposite of **auxotroph**.

Punnett square a table used to calculate the chance of different genotypes appearing from a genetic cross, when the genotypes of the parents are known.

pure breeding an individual or population which reproduces sexually but remains genetically unchanged from generation to generation: in some cases a synonym for homozygous.

R factor see **resistance factor**.

radioactive labelling as with isotopic labelling – using radioactive isotopes; see **isotopic labelling**.

rapid lysis mutant a mutant form of bacteriophage producing unusually large, clear plaques.

recessive of an allele: not able to show its effect on phenotype in the presence of a dominant partner; see **dominant**.

reciprocal cross a cross which differs from its reciprocal only in which genotype is the female parent and which the male parent.

recombinant DNA a molecule formed by the joining of two pieces of DNA from different sources.

recombinant DNA technology the techniques associated with the production and manipulation of recombinant DNA: a form of genetic engineering.

recombination the breakage and reunion of two pieces of homologous DNA or two chromatids, such that genes are put together in new combinations.

reduction division see **meiosis**.

renaturing the restoring of the double stranded structure to DNA which has been denatured.

replication see **DNA replication**.

replication fork the place where a DNA molecule is in the process of replicating; the two strands of the DNA separate prior to replication and the two arms of the fork are formed by the separating strands.

reproduction an increase in the number of individuals; reproduction may or may not be associated with a sexual process.

repression see **enzyme represssion**.

repressor a substance which inhibits the action of a gene or genes, often the end product of a biochemical pathway controlled by the gene or genes in question.

resistance factor a bacterial plasmid carrying genes for antibiotic resistance.

resolution the ability of an optical system to distinguish two objects as separate; the smaller the distance apart of two such objects, the greater the resolution.

restriction endonuclease an enzyme which cuts DNA into discrete pieces.

restriction enzyme an alternative name for **endonuclease**; also called **restriction endonuclease**.

restriction fragment a piece of DNA obtained from the action of an endonuclease.

restriction site a particular sequence of bases in DNA which can be cut by a restriction enzyme; restriction endonucleases only cut DNA at certain sites where such sequences are recognised.

retrovirus a type of virus which contains RNA as its genetic material and has the ability to convert RNA to DNA by **reverse transcription**.

reverse transcriptase the enzyme employed by retroviruses to synthesise DNA from an RNA template.

reverse transcription the process of converting RNA to its complementary DNA: the reverse of the synthesis of mRNA.

ribose the five carbon sugar component of RNA.

ribonucleic acid RNA: the type of nucleic acid containing ribose.

ribosome the cell organelle that is the site of protein synthesis.

RNA ribonucleic acid.

^{35}**S** radioactive isotope of sulphur.

segregation the separation, one into each meiotic product, of a pair of alleles.

selection a force that results in a change in the frequency of a particular allele in a population by influencing the reproductive capacity of individuals carrying that allele.

selective advantage the likelihood that a certain genetic type will reproduce more successfully than other types.

selective medium a bacteriological medium which only allows the growth of certain types of microorganism, allowing these types, therefore, to be selected.

semi conservative replication replication, as of DNA, which produces two copies, each of which contains half the original.

sex the process that involves the mixing of genetic material from different cells; in higher organisms sex invariably involves the production of gametes and the subsequent fusion of gametes.

sex chromosome a chromosome that influences the sex of an individual and which is not entirely homologous with its partner; usually sex chromosomes are either X or Y chromosomes and the two sexes are XX and XY respectively.

sex linkage the linkage of genes to the sex chromosomes.

sexual dimorphism there being two distinct forms, male and female.

shotgunning a technique for isolating a particular gene from a sample of DNA broken into fragments by restriction enzymes.

sickle cell anaemia a hereditary disease of humans affecting the red blood cells.

somatic cell any cell that is not part of the germ line.

sperm a mobile male gamete typical of animals.

spontaneous mutation a change in the genetic material brought about with no obvious cause.

sticky end the partly single stranded end of a piece of DNA produced by cutting with a restriction enzyme; the sticky end has a few exposed nucleotide bases inviting base pairing.

T-even phage one of a series of bacteriophages given the notation T followed by an even number.

T-one phage the first numbered of the T series of bacteriophages.

temperate phage a bacteriophage which, having infected a bacterial host, fails to reproduce and lyse the cell but, rather, becomes a prophage and renders the host lysogenic.

template a mould allowing the production of a copy according to a pattern: one strand of DNA is a template for mRNA synthesis or for the synthesis of its complementary strand during DNA replication.

terminator codon one of three codons: UAG, UAA or UGA which do not code for any amino acid but, instead, act as full stops in the genetic code and terminate the process of linking amino acids together to form a protein.

test cross a cross, usually a back cross, designed to test the genotype of an individual.

tetrad (a) the four chromatids paired at prophase and metaphase of first meiotic division, (b) the four products of a single meiotic division, (c) the four spores or pairs of spores in an ascus.

tetrad analysis analysis of the genotypes of the four immediate products of a single meiosis in order to interpret recombination events.

tetraploid having four complete sets of chromosomes.

thymine one of the bases of DNA: thymine pairs with adenine.

thymine dimer a structure formed by the combination of adjacent thymine bases in DNA under the influence of ultra violet light; the formation of thymine dimers is extremely damaging to DNA function.

totipotency the idea that even differentiated cells of a multicellular organism contain the whole genetic plan for the organism.

transcription the conversion of the genetic code in DNA into the equivalent code in mRNA: mRNA synthesis.

transduction the transfer of genes from one bacterial cell to another by means of a phage vector.

transductional mapping mapping bacterial genes using the principle that if a phage vector carries two genes together (**cotransduction**) they must be very closely linked.

transfection infecting a bacterium or other cell with a virus using pure DNA only.

transfer RNA the type of RNA whose job it is to attach specifically to one particular amino acid and carry it to the site of protein synthesis; each transfer RNA has an anticodon complementary to the mRNA codon for a particular amino acid.

transforming principle now known to be DNA: the component of organic systems capable of bringing about genetic transformation.

transgenic organism an organism altered by genetic engineering; also called **genetically modified (GM)**.

translation the conversion of the genetic code in mRNA into protein.

translocation a mutation of a chromosome in which part of a quite different chromosome is transferred to it.

trinucleotide a short length of DNA containing three nucleotides.

triplet an abbreviated way of referring to the three bases that make up a codon or anticodon.

triplet binding assay a method by which the genetic code was elucidated: in the presence of a trinucleotide (**triplet**) and ribosomes, only one type of tRNA will bind to the other two, the type that carries the amino acid coded for by that particular triplet.

triploid having three complete sets of chromosomes.

tRNA transfer RNA.

trisomy having three copies of one chromosome rather than a pair.

tryptophan synthetase the enzyme responsible for the synthesis of the amino acid tryptophan.

ultraviolet light light of wavelength just beyond the visible violet end of the spectrum; UV light is damaging to DNA and causes mutations.

unicellular of organisms made of a single cell.

uracil one of the bases of RNA: uracil pairs with adenine.

vector literally a carrier: in genetic engineering a carrier of DNA, such as a plasmid or a virus (**cloning vector**).

virus a non cellular biological parasite consisting of a protein coat and a core of nucleic acid.

wild type having no mutant genes: the most common form of an organism; often meaning being homozygous for a dominant allele.

X chromosome one of the sex chromosomes, the other being Y.

X linkage of genes that are linked to the X chromosome and have no counterpart on the Y chromosome.

X ray diffraction a technique for analysis of molecular structure.

Y chromosome one of the sex chromosomes, the other being X.

Y linkage of genes that are linked to the Y chromosome and have no counterpart on the X chromosome.

Zygote the cell formed by the fusion of two gametes: if the gametes are haploid the zygote will be diploid.

Web Sites about Genetics

A Level Biology
http://www.biology.demon.co.uk/Biology/mod2/mendel/genetics.htm
A site for A-level students, with lecture notes on all aspects of genetics.

Access Excellence
http://www.accessexcellence.org/
A place in cyberspace for biology teaching and learning. This site has a search facility and also gives a number of web site links.

Ageing and Genetics
http://www.biorap.org/tg/tgagecontents.html
This is all about ageing and genetics and is an excellent source of general as well as specifically related genetic information. Lesson 2 in the teachers' guide is on basic genetics.

American Society of Human Genetics
http://www.faseb.org/genetics/askg/ashgmenu.htm
This contact site includes a section of press highlights relating to human genetics.

The Biology and Educational Resource Center
http://schmidel.com/bionet.htm
This site also gives links to other useful web sites.

Bock Labs
http://www.bocklabs.wisc.edu/ed/virtut1.html
A primer on molecular virology with good background knowledge.

Disease genetics
http://raven.umnh.utah.edu/review/redo/diseasemenu.html
An electronic module for learning about how genetics relates to disease.

Electronic Scholarly Publishing
http://www.esp.org/site
ESP is dedicated to the electronic publishing of scientific and other scholarly material, especially the history of science, genetics, computational biology and genome research. Of particular value at this site is a chronology giving a historical context to genetics.

Genetic Diseases
http://www.ben2.ucla.edu/~vpalchev/disease.htm
Another site dealing with genetic diseases. This one covers a lot of basic background and answers frequently asked questions about genetic diseases.

Human Genetics Advisory Commission
http://www.dti.gov.uk/hgac/
The HGAC deals with issues such as embryo research, genetic testing, cloning and genetic modification.

Medical Genetics Centre
http://www.genecare.com/dna.html
Contains information on genetic testing for conditions such as cystic fibrosis and paternity disputes.

Mendel Web
http://hermes.astro.washington.edu:80/mirrors/MendelWeb/
An educational resource for anyone interested in the origins of classical genetics. It includes Mendel's original paper and answers to frequently asked questions.

Molecular Genetics
http://gened.emc.maricop.edu/bio/bio181/BIOBK/BioBookDNAMOLGEN.html
This is the web address for one chapter (on DNA, protein synthesis and molecular genetics in general) of a collection of excellent resources on aspects of biology and will lead to other chapters on, for example, introduction to genetics, gene interaction and human genetics. It is very well illustrated.

National Cancer Institute
http://cancernet.nci.nih.gov/h_detect.html
Site for the National Cancer Institute. Go to genetics, then click on understanding gene testing. You will find answers to basic questions such as what is a gene, plus an excellent glossary. This site is a mine of information.

The New Genetics
http://arginine.umdnj.edu/~swartz/teachgen.html
The best of the lot: this is perhaps the only web site you'll need. It's called a resource for teachers and students and gives you a menu taking you to numerous other very valuable sites.

Primer on Molecular Genetics
http://www.ornl.gov/hgmis/publicat/primer/intro.html
A primer on molecular genetics from the US Department of Energy. It contains plenty of information (very much like a book).

USA Today
http://www.usatoday.com/life/health/lh118.htm
A site for Life magazine giving news of gene therapy.

Virtual Fly Lab
http://vflylab.calstatela.edu/edesktop/VirtApps/VflyLab/Design.html
The Virtual Fly Lab allows the user to set up crosses between two genetic types of fruit fly and obtain results that comply with the expected inheritance pattern of the chosen genetic variables.

* Note: more genetics links for students are freely available at the Studymates web site:

http://www.studymates.co.uk

Index

Genetics